FLOODS, DROUGHTS, AND CLIMATE CHANGE

The University of Arizona Press
© 2002 The Arizona Board of Regents

⊚ This book is printed on acid-free, archival-quality paper.
Manufactured in the United States of America
07 06 05 04 03   6 5 4 3 2

Library of Congress Cataloging-in-Publication Data
Collier, Michael, 1950–
Floods, droughts, and climate change / Michael Collier and
Robert H. Webb.
p. cm.
Includes bibliographical references and index.
ISBN 0-8165-2250-2 (pbk. : alk. paper)
1. Climatology.   2. Climatic changes—Environmental aspects.
I. Webb, Robert H.   II. Title.
QC981 .C677   2002
551.6—dc21
2002005314

British Library Cataloguing-in-Publication Data
A catalogue record for this book is available from the British Library.

FLOODS, DROUGHTS, AND CLIMATE CHANGE

Michael Collier and Robert H. Webb

The University of Arizona Press

Tucson

To Rose and Toni

Contents

Illustrations

All photographs are copyright Michael Collier unless otherwise noted.

Acknowledgments

The authors wish to particularly thank Michael Dettinger and Dan Cayan for their input and encouragement throughout preparation of this book. Many of Dettinger's comments found their way into the finished manuscript. Kelly Redmond, David Enfield, Richard Hereford, and Ed Peacock provided thoughtful commentary that helped shape our thinking along the way. Diane Boyer and Peter Griffiths helped us with the photographs and illustrations, respectively.

Walter Nemanishen and Andrew Cullen generously led us through the prairie farms of southeastern Alberta. Clara Bravo and Michael White shared their knowledge of Trujillo and the surrounding communities in northern Peru. Kenneth Rivera, assistant director of Honduran water resources, and Caronte Rojas guided us through the aftermath of Hurricane Mitch in Tegucigalpa. Peter Hearne with USAID and Osvaldo Munguia and Adelberto Padilla with MOPAWI helped make our visit to Honduras possible. Captain Uli Demel and the Explorer Shipping Company provided safe and spectacular passage to the Antarctic Peninsula and back.

Special thanks to Kevin Trenberth, Tim Barnett, Henry Diaz, Alexander Gershunov, Gary Sharp, Curtis Ebbesmeyer, Nick Brooks, and Jeff Dean, who were willing to spend time going over the authors' ideas. Phil Turnipseed, Brian Bell, Henry Noble, Vic Baker, William Brown, Jeff Phillips, and Curt Storlazzi were kind enough to field individual questions.

Barry Lopez provided early inspiration to look at this subject in its widest and most human terms.

Abbreviations

ENSO   El Niño–Southern Oscillation
HYF   hundred-year flood
ITCZ   Intertropical Convergence Zone
NOAA   National Oceanic and Atmospheric Administration
PDO   Pacific Decadal Oscillation
QBO   Quasi-Biennial Oscillation
SOI   Southern Oscillation Index
TOGA   Tropical Ocean Global Atmosphere Project
USGS   U.S. Geological Survey

FLOODS, DROUGHTS, AND CLIMATE CHANGE

Entering the Atmosphere

One would think that the atmosphere —all 5,000 million million tonnes of it[1]— would be more conspicuous. We spend our lives within this sea of air. It permeates every void and lodges itself between every hair. We breathe it, we walk through it, and yet the atmosphere remains a mystery, invisible to us. Are fish aware that they swim in water? The atmosphere is so intimate, yet it slips like mist through our fingers. Without the extremes of wind and rain, heat and cold, we would scarcely even acknowledge the atmosphere's existence.

If we are to understand why floods and droughts occur, we must begin with the air we breathe. Climate is the fundamental link between the two. Hippocrates wrote about the effects of climate on human health twenty-four hundred years ago. Aristotle expounded on patterns of weather fifty years later. The Greeks were aware of orderly zones of climate; indeed the word derives from κλιμα, which refers to the gradational effects of a slope. Despite the Greeks' conjectures, subsequent observers could no more comprehend atmospheric dimensions than they could grasp dreams vaguely remembered upon waking.

It wasn't until people thought to measure temperature, pressure, and motion that air in some sense became tangible. Santorre Santorio invented the thermometer in 1612, and Evangelista Torricelli built a barometer in 1643. With these tools, Robert Boyle was able to demonstrate his law of gases in 1661, which explained the relationships between pressure, volume, and temperature. Edmund Halley drew the first map of trade winds seven years later. With these maps, George Hadley offered an elementary understanding of global atmospheric motion in 1735 that stands as the basis for much of global climatology today.

The study of the atmosphere has advanced dramatically over the last one hundred years. During the First World War, Norwegian meteorologists described embattled warm or cold bodies of air separated by "fronts." The jet streams were discovered during the Second World War by high-altitude pilots who were impeded while trying to make their way westward over the Pacific Ocean. In 1957 the International Geophysical Year focused scientists' attention on the global environment. Since then, a coherent system of oceanic and atmospheric measurement stations has gradually been developed. Tiros I was launched in 1960, the first weather-observation satellite to be put into orbit. Today the Tropical Ocean Global Atmosphere Project (TOGA) network of monitoring buoys collects information over large portions of the world's tropical oceans.

The National Oceanic and Atmospheric Administration (NOAA) and similar institutions worldwide have continuously gathered objective meteorologic data for decades: wind, temperature, precipitation, barometric pressure, and atmospheric moisture. Data gathered from the oceans, atmosphere, and space are the foundation for an understanding of the interconnectedness of our evolving atmosphere. All of these tools have sharply focused our vision of atmospheric processes. Indeed, only in the last third of the twentieth century did satellite imagery allow us to comprehensively see the awesome and complete fury of a hurricane.

From one day to the next, we're touched by the vagaries of weather: extremes of heat or cold, rain or drought. Storms break and rivers burst their banks; or skies remain relentlessly clear for months as we watch animals die and reservoirs dry up. Over the years, season after season, one learns what to expect from the weather. With time, the cumulative experiences of storm, wind, and heat coalesce into a sense of climate, an important component of our sense of place. Although this "sense of place" is invaluable to us as individuals, it doesn't offer scientists or society much of a foundation for the broader picture of a global climate, much

A flash flood on a wash in the Tucson Mountains in Arizona following a summer monsoon thunderstorm, 1982.

less any insight into how climate might be changing or how it affects our lives. For that we need objective data from worldwide monitoring systems.

For more than a century, the U.S. Geological Survey (USGS) has assessed river flows throughout this country. USGS scientists have maintained river gaging stations since 1895, amassing long-term records of water flow and sediment transport. These records provide an indispensable backdrop against which one can evaluate "aberrant" conditions of flooding along a particular stretch of river or drought within a given river basin. Hydrologists and climatologists have long been aware of the role of regional climate in the prediction of floods or in understanding drought. With our growing sense of a variable climate, it's appropriate to reassess these concepts of flood and drought, not as isolated events, but as phenomena connected on a worldwide scale.

This book examines floods and droughts within the wider context of climate and climate change. It introduces the concepts of global weather, puts these processes into the longer-term framework of climate, and explores the idea that patterns of climate evolve through time. With this foundation, it becomes increasingly clear that floods and droughts, once considered isolated acts of God, are often related events driven by the same forces that shape the oceans and the entire atmosphere.

Cutting Edges of Climate

Extremes of weather shape us as individuals and as a society. Storms and searing heat sharpen the edges of our culture; floods and droughts hone the edges of our landscape. The hurricane of September 1900 swept six thousand people from the streets of Galveston, Texas.[1] In 1927 close to a million people—almost 1 percent of the U.S. population—left their homes to huddle atop levies from Illinois to Louisiana as the Mississippi River flooded a combined area equal to Massachusetts, Connecticut, New Hampshire, and Vermont.[2] Six years later, winds of the Dust Bowl began to inhale 100 million acres of western Kansas, eastern Colorado, and the panhandles of Oklahoma and Texas; the summers of 1934 and 1936 were the hottest since the turn of the century.[3] John Steinbeck forever focused this catastrophe in the American consciousness when he wrote *The Grapes of Wrath* in 1939:

> And then the dispossessed were drawn west—from Kansas, Oklahoma, Texas, New Mexico; from Nevada and Arkansas, families, tribes, dusted out, tractored out. Car-loads, caravans, homeless and hungry; twenty thousand and fifty thousand and a hundred thousand and two hundred thousand. They streamed over the mountains, hungry and restless—restless as ants, scurrying to find work to do—to lift, to push, to pull, to pick, to cut—anything, any burden to bear, for food. The kids are hungry. We got no place to live. Like ants scurrying for work, for food, and most of all for land.

Floods and droughts are not aberrations. Floods are initially more conspicuous than droughts because they can occur over days or weeks instead of months or years. Droughts require a more persistent weather pattern before they are recognized. We commonly assume they are end members of the spectrum of possible meteorologic conditions for a given locale. Another emerging perspective, however, is that floods and droughts arise from conditions that are somehow different than the established norm. *Climate* may not turn out to be a smooth continuum of meteorologic possibilities after all, but rather the summation of multiple processes operating both regionally and globally on differing time scales.

Floods and droughts are neither random nor cyclic.[4] They may seem that way when our noses are pressed against the windowpane of the present, but they aren't. To the Dakota farmer whose lips were cracked dry during the Dust Bowl, it seemed that rain would never sweep in over the horizon. To the Honduran peasant whose home and family have just been swallowed by the hurricane-swollen Río Choluteca, dry blue skies are the vanishing memory of a former life. Up close, it's all but impossible to see the true patterns of weather.

The extremes of flood and drought occur within the context of climate, a context that is both local and global. One must understand the geography and meteorologic response of a given watershed to understand its history of flooding. One should also look beyond basin boundaries to appreciate the coherent patterns that influence weather regionally.[5] Almost fifty years ago, Jerome Namias suggested that local droughts can be the manifestation of anomalous patterns of atmospheric circulation arising from changing ocean-surface conditions half a world away.[6] The same is true for floods.

Droughts

Drought is more than a simple lack of rainfall. Drought is a persistent moisture deficiency below long-term average conditions that, on average, balance precipitation and evapotranspiration in a given area. Not all droughts are created equal; similar moisture deficits may have very different consequences depending on the time of year at which they occur, pre-

A ranch near Gila Bend, Arizona, during the 1993 floods on the Gila River. These floods, part of the 1992–93 El Niño event, were the largest to affect the entire state since 1891.

existing soil moisture content, and other climatic factors such as temperature, wind, and relative humidity. Drought can be defined in terms that go beyond the meteorologist's rainfall measurements. *Hydrologic drought* occurs when surface water supplies steadily diminish during a dry spell. If dry conditions continue, groundwater levels could begin to drop. *Agricultural drought* occurs when a moisture shortage lasts long enough and hits hard enough to negatively impact cultivated crops. Soil conditions, groundwater levels, and specific characteristics of plants also come into play in this functional definition of drought. *Ecologic drought* is detrimental to native plants that don't have the benefit of irrigation.

The dryland farmers of southeastern Alberta know about several kinds of drought, particularly agricultural drought. John Palliser had explored that country from horseback for the Canadian government during the dry years of 1857–59.[7] He warned that southeastern Alberta and southwestern Saskatchewan would not reliably sustain human (i.e., agricultural) life. The land was fundamentally arid, receiving 400 or fewer

Tracks and silos at Hanna, Alberta, 1998.

millimeters (mm) of rain a year; some years there might as well be no rain at all. Fifty years later, Palliser's admonitions were swept away by the newly arrived Canadian Pacific Railway as it launched one of the most far-flung publicity campaigns ever undertaken.

In 1901, 600,000 hectares within the "Special Areas" (defined by a dryland ranch life that would all but collapse thirty-five years later) had seventy-five residents.[8] Fifteen years of rain and railroad boosterism followed. By 1916, twenty-four thousand people were busting sod on their quarter-section homesteads. Throughout the larger Palliser Triangle of southeastern Alberta and southwestern Saskatchewan, 280,000 people were farming almost 3 million hectares before the First World War.[9] About 100,000 hectares within the Special Areas would be broken by the plow by 1924.

Harry Gordon's parents left Calgary and joined the land rush to the prairies around Hanna, Alberta, in 1913; Harry was one month old. Like their neighbors, the Gordon family thrived during the wet years through 1915. A lucky farmer could expect 100 bushels to the hectare. Wheat

An abandoned home near Dorothy, Alberta, 1998.

prices soared as World War I raged. Then the rain ceased. Drought sucked
the Gordon farm dry from 1917 through 1926. In nearby Medicine Hat,
1 hectare of land planted in wheat produced 17 bushels in 1917, a couple
bushels in 1918, and none in 1919. Harry learned to do without; later he
avoided debt like the plague itself. He learned to always leave a year's
worth of grass on the range for his cattle.

Times improved on the Alberta prairie during the mid-1920s, but
drought conditions returned during the Dust Bowl years. Sixty-six thousand people gave up their plows from 1931 through 1936. The populations
of towns within the Special Areas shrank by 40, 60, or even 80 percent.
Canada's federal government created the Prairie Farm Rehabilitation Administration in 1935 to assist in the formation of irrigation districts. The
Special Areas were officially designated in southeastern Alberta to assist
dryland farmers who struggled to stay on their land. Harry Gordon hung
on through good times and bad. He learned to live with drought.

Southeastern Alberta is drought prone, not because of its low average annual rainfall, but because its climate departs from average rainfall

Windblown sand encroaching on Lorne Coles's farm near Hanna, Alberta, during drought conditions associated with the 1997–98 El Niño.

patterns frequently enough to hamstring people who try to live and farm there. South of Hanna, Henry Haugen raises cattle around Manyberries, near the Alberta/Montana border. His grandparents settled on this land in 1900; he was born just a kilometer up the road. Rainfall here averages 350 mm a year; in some years, it's less than 200 mm. Henry has run as many as four hundred cattle. Typically, each cow needs 200 hectares of grazing pasture. His well produces water from a depth of 100 meters (m) with an extremely high salt content: dissolved solids total more than 3,000 parts per million. The Haugens hung on through the 1930s by abandoning any expectation of return on their investment; income was plowed back into expenses year after year. That solution may not always be good enough for his three sons who continue to ranch with him. If water drops any lower in the well or if his sons' wives become any less content, one by one they will leave to live elsewhere.

Southwest of Alberta, in the Cascade Mountains of Oregon, Barry Lopez pondered the McKenzie River alongside his home as it dropped with drought one year.[10]

I awoke one night and thought I heard rain—it was the dry needles of fir trees falling on the roof. . . . I have trained myself to listen to the river, not in the belief that I could understand what it said, but only from one day to the next know its fate. . . . It was in this way that I learned before anyone else of the coming drought. Day after day as the river fell by imperceptible increments its song changed, notes came that were unknown to me. . . . I would say very simple prayers in the evening, only an expression of camaraderie, stretching my fingers gently into the darkness toward the inchoate source of the river's strangulation. I did not beg. There was a power to dying, and it should be done with grace. I was only making a gesture on the shore, a speck in the steep, brutal dryness of the valley by a dying river.

It was drought, along with war and famine, that biblical people feared most. We have spent the last two thousand years trying to anticipate the rhythms of climate, trying to understand the machinations of rain. But these rhythms are elusive. Half a century ago, in a lovely book entitled *Drought, Its Causes and Effects*, I.R. Tannehill sighed as he observed, "In conclusion, the writer regrets that it is necessary to say that the problem of drought is not completely solved."[11] We continue to search for an underlying cause for drought. Lorne Coles ranches near Hanna. He stands alongside a pasture fence with a handful of dry sand sieving though his fingers. Lorne and his dryland neighbors in southeastern Alberta are curious to know what we find.

Floods

The cadence of wet and dry years is difficult to comprehend, let alone predict. A tantalizing pattern links drought and its inverse, flooding, in many places around the world. Floods in one location and drought in another—so obviously different—often exist within similar global configurations of climate.

Flood and drought: Both are etched deeply into the collective human experience. Flooding on the Huang Ho River took the lives of 3.7 million Chinese in 1931.[12] Drought in Africa's Sahel Desert claimed at least

150,000 lives during the second half of the twentieth century. Over time, drought has been more deadly than flooding. Drought depends on the persistence of dryness over months or years. Catastrophic floods can explode suddenly out of a single summer thunderstorm. Flooding, however, can also be caused by a months-long buildup of moisture, such as the fast melting of a heavy winter's accumulation of mountain snow or soil saturated by high seasonal rainfall. All floods, of course, are shaped by the basin through which they flow.

Typical springtime floods on the Mississippi River have peak discharges on the order of 30,000 cubic meters per second (m^3/s). On March 21, 1997, the Mississippi River at Vicksburg, Mississippi, was flowing at twice that volume: an irresistible python of water 2.5 kilometers (km) wide and 40 m at its deepest, coursing past the delta countryside at 10 kilometers per hour (km/hr). Floods on the Amazon River are typically ten times as large as those on the Mississippi, but to find truly big floods one must travel surprising distances through time or space.

During the late Pleistocene, about 15,000 years ago, the breakup of ice dams on the now-vanished Lake Missoula triggered a series of floods that leveled everything in their path from Montana to the Pacific Ocean. Peak discharge of one of those floods—the Rathdrum Prairie event—has been calculated at 20 million m^3/s, that is, a thousand Mississippi River floods flowing together.[13]

Outwash channels on Mars—complete with streamlined uplands, longitudinal grooves, dry waterfalls, scour marks, and fan-shaped deposits of sediment—are 100 km wide and 2,000 km long. After guessing at a multitude of parameters, including channel slope and sediment load, scientists have estimated that the peak flow volume of these Martian floods was 300 million m^3/s.

We need to look only as far as the earth to see the various mechanisms that trigger floods. Earthquakes and landslides that occur beneath an ocean floor can induce *tsunamis*. These are waves that flash unnoticed across an entire ocean to pile up against an unsuspecting coastline half a world away. Dams, either man-made or natural (such as ice jams or landslides), can burst. Hurricanes and cyclones whip up tidal surges that inundate coastal lowlands. Rivers can swell in response to precipitation. Tsunamis and dam failures have certainly wreaked havoc through the ages. In 1960, when Chile was rocked by the greatest earthquake ever recorded (magnitude 9.5), the associated tsunami crossed the Pacific in twenty-two hours, drowning 5,700 people in Hawaii and Japan.[14] After the La Josefina landslide blocked the Paute Valley near Cuenca, Ecuador, in 1993, its catastrophic failure released a flood that approached 10,000 m³/s.

Of the four flood mechanisms listed above, the last two—tidal surges and especially river response—are most intimately tied to the whims of weather. It is on these two that we will focus. The Ganges-Brahmaputra-Meghna River system can swell dramatically when the Indian monsoon is especially torrential. In September 1987, the lives of 25 million people in Bangladesh were disrupted when monsoonal flooding inundated 2 million hectares of cropland. These floods were anticipated by days if not weeks; even so, 5 million homes were destroyed and 2,000 people died. But what the Bangladeshi truly fear is flooding that sweeps in, not from the mountains, but from the Indian Ocean. Tidal surges are domes of water 75 to 150 km across that are driven aground in front of an incoming storm. Surges from just two tropical cyclones killed 440,000 people in Bangladesh in November 1970 and April 1991.[15]

Floods are most likely to kill when they occur without warning. Residents of Nelson, Nevada, awoke to blue skies on the morning of September 14, 1974. As the day progressed, convective clouds built over the mountains behind town. A single thunderstorm lingered over the upper end of the 59 km² basin leading into El Dorado Canyon. Sluggish upper-level winds carried the cell slowly eastward toward the canyon's mouth, so that rain kept falling on the moving flood crest, amplifying floodwaters that had already begun to rush through the canyon.[16] More than 2,000 m³/s swept down the usually dry wash, offering no more than minutes of warning to the people below. Nine drowned. The tiny community of Nelson vanished into Lake Mohave on the Colorado River.

In contrast, the Mississippi River's great flood of 1993 had its beginnings in the upper basin's gradual accumulation of above-average rainfall as early as July 1992. Warm waters in the Pacific Ocean had reinforced a ridge of high pressure over the western United States. The subtropical jet stream was enhanced, adding kinetic energy and moisture to the flow

Flooding on the Mississippi River near Cape Girardeau, Missouri, during the summer of 1993. This flooding was caused by well-above-average winter and spring precipitation in the upper Mississippi River drainage associated with the 1992–93 El Niño.

of air coming into the Midwest. Then during the summer, a ridge of high pressure became embedded above the East Coast. Clockwise rotation of wind around this ridge aimed a fire hose of moist air northbound from the Gulf of Mexico against the Midwest.

A river of atmospheric moisture was flowing from the Gulf of Mexico into the central and eastern United States. The westerly jet stream served as the match that ignited this fuel flowing in from the gulf. Embedded within this moist regional air mass were at least 175 cores of intense thunderstorm activity, each locally producing 150 mm or more of rain. Thunderstorms popped off like a string of firecrackers all summer long. Weekly average rainfall throughout the upper Mississippi and Missouri River basins from June through August 1993 was 28 billion m³. After forty weeks of this deluge, the precipitation total was twice the volume of Lake Erie.[17]

Minor flooding lapped against levees at St. Louis for forty-four days in April and early May 1993. The Mississippi and Missouri Rivers rose, but it wasn't until June 20 at Hatfield, Wisconsin, that the first of a thousand levees broke.[18] From that point on, the rivers rose inexorably, each day surpassing the previous day's estimate of flood height. Locks were flooded; the rivers were closed to barge traffic; grain prices soared. At St. Louis, the Mississippi River finally crested at 15.1 m on August 1, 1993, with a discharge of more than 28,000 m^3/s. Record flooding was observed along 2,900 km of river within the upper basin, and major flooding occurred on an additional 2,000 km. By the end of August, 50,000 people had been displaced and almost 40,000 km^2 of prime cropland were covered by floodwaters. Fifty-two people died. Losses totaled $18 billion.[19]

Floods don't just happen because it rains. Floods happen because rain falls on saturated ground, because warm rain falls on an existing snowpack, because rain falls heavily throughout an entire basin, or because the basin has been changed (either naturally or otherwise) so as to retain or heighten floodwaters that would have otherwise rolled on through without making a mess. Floods are usually more localized than droughts—both in time and in space—because floods require these specific preexisting hydrologic and meteorologic conditions.

Spring melting of winter snow is always a time of high river flow. In many regions, river channels shape themselves to accommodate these annual events, building banks and gradually raising floodplain terraces in response to each year's high water. Meltwaters will usually be released from the snowbound mountains in an orderly progression as springtime temperatures begin to rise. Floods will occur if temperatures rise faster than expected or if rain falls on snow that is already near its melting point.

Destructive flooding doesn't always have to follow high precipitation. The mountains of western Washington and Oregon were buried beneath 30 m of snow in the winter of 1998–99, far more than usually falls. Communities along the Northwest Coast held their breath as spring arrived, waiting for the floods that seemed inevitable. But nothing out of the ordinary happened. Because the spring turned out to be relatively cool, the snowmelt occurred in an orderly fashion and rivers remained within their banks. Tallying precipitation does not necessarily equate with flood forecasting.

A Rio Nido, California, resident inspecting damage from the February 1998 debris flow associated with the 1997–98 El Niño.

During the winter of 1997–98, California was inundated by a series of storms that broke records for precipitation throughout the state. Eureka received rain on fifty-two of the fifty-nine days of January and February. San Francisco registered 508 percent of normal precipitation during the month of February. The people of Rio Nido along the Russian River in the central Coast Ranges of California were blasted awake on February 7 by a debris flow that reduced homes to matchsticks along Upper Canyon Three. The community waited without defense, holding its breath 600 feet beneath a saturated hillside that threatened to erase 140 homes that were at risk if the hillside slipped. Pescadero Canyon was repeatedly blocked by mud slides that had taken lives at Loma Mar in the mountains above Palo Alto. Homes in Los Angeles were transformed into nightly news icons as foundations eroded and their walls gradually disappeared.

Floods are fickle, requiring very specific conditions beyond merely wet weather. A flood might hit one basin yet inexplicably ignore another

Damage in a subdivision caused by a debris flow on the Rio Nido, a tributary of the Russian River in the Coast Ranges of California. This debris flow occurred during the intense winter El Niño rains of February 1998.

basin nearby. Sometimes whole towns or villages are wiped out; sometimes the damages seem more like the luck of the draw. Whether widespread or local in scale, floods are set up by large-scale atmospheric processes that are in some ways the inverse of droughts. Whether the flood occurs or not when those conditions are present is another matter that befuddles flood forecasters.

The Weather 3 Machine

We experience our immediate environment moment by moment. Now it's hot; now it's cold. The slightest puff of wind can suddenly make a muggy summer day more comfortable. In winter, dry air can be hard, even brittle. Taken over a period of weeks to months, these experiences add up to local *weather*. The weather this winter in Seattle might tend toward rain, while the spring weather in Tucson is likely to be sunny. Weather is what we're experiencing now.

We are barraged daily—indeed every few minutes—by the latest-breaking, up-to-date, Doppler-enhanced, color-spangled weather report. From our human perspective of three score and ten years, it's not easy to achieve a longer view of climate; we are better at licking a finger, sticking it up in the wind, and appreciating weather as it occurs.

How does one achieve a tangible understanding of the processes of weather? Global atmospheric circulation is frighteningly complex, whether observed on a large or small, brief or long-term basis. This complexity exists because atmospheric flow is volatile. Small disturbances—a storm here, a hot spot there—can explosively amplify to near-global scales and then merge with more stable, typical flows. The atmosphere is not a quiet pool; it is a surging, roiling river. And like a river, the atmosphere has well-worn pathways that it travels. These pathways create our *climate*. An understanding of weather patterns requires an awareness of turbulent movement in time and three dimensions. Don't be distracted or dismayed by this complexity. Simplified models can illuminate fundamental processes, providing insights into atmospheric processes.

Climate is all about thermodynamics, because heat drives the earth's weather machinery. The greatest portion comes from the sun, though a much smaller fraction is still being released from nuclear decay deep within the earth. The sun converts 600 million tonnes of hydrogen into helium every second. A little less than 1 percent of that hydrogen (a mere 4 million tonnes per second) is converted into energy rather than helium. Solar energy reaches us as radiated electromagnetic waves broadcast across a 150-million-km void of space. Our planet, a mere 12,000 km across, catches only an infinitesimal fraction of these waves and turns them into heat.

Solar radiation is produced in a range of wavelengths, most of which are between 0.05 to 2.0 micrometers (μm).[1] Most of the solar radiation that reaches the earth is in wavelengths of 0.4 to 0.7 μm, which happen to be the visible spectrum of light to which our eyes have been adapted by evolution. Radiation between 0.01 and 0.4 μm, which is ultraviolet, is absorbed by oxygen and ozone in the upper atmosphere. Radiation from about 0.7 to 2.0 μm, which is infrared, is absorbed by water molecules in the atmosphere. The lion's share of the radiation that penetrates the filter of the atmosphere and heats our little planet is the visible spectrum.

The earth receives only one part in two billion of the sun's total radiated energy: 1.94 calories received over each square centimeter (cm^2) of its surface each minute, if that patch of the earth happens to be oriented at right angles to the sun.[2] When this energy strikes the earth's atmosphere, two things happen: Some solar radiation is reflected back into space, while the remainder is absorbed by the earth and atmosphere and then converted into heat. Reflectivity *(albedo)* varies significantly depending on the angle at which the radiation arrives and on the color and roughness of the surface it hits. Wear a black shirt and feel warm; wear a white shirt and stay cool.

Overall, the earth and its atmosphere reflect 31 percent of all incoming solar radiation back into space. Sunlit clouds reflect between 40 and 90 percent, depending on their thickness and internal structure. Various land surfaces will reflect anywhere from 5 percent (croplands) to

90 percent (fresh snow). Sun rays that strike the ocean perpendicularly around the equator are almost entirely absorbed, while rays glancing off the ice and cold waters of the polar regions are almost entirely reflected. Of course, some reflected radiation hits clouds and bounces back to the earth, complicating the energy balance.

Some incoming radiation is absorbed by the molecules that compose our atmosphere, making our planet a nicer place to live. Nitrogen, oxygen (as either O_2 or O_3), and water vapor constitute 99 percent of the atmosphere. Nitrogen (as N_2O) absorbs only a small amount of incoming solar energy. Oxygen optimally absorbs wavelengths shorter than 0.4 μm and thus plays a minor role in capturing solar radiation. Ozone (O_3) resides high in the stratosphere at altitudes of 30–80 km and is particularly adept at absorbing incoming ultraviolet rays that would otherwise be damaging to life on the earth.

Just over half (58 percent) of the sun's incoming energy actually reaches the earth's surface.[3] A portion of this is immediately reflected from the surface back into the atmosphere and out into space. The rest is absorbed by the oceans and landmasses. Now the real fun starts: the earth itself begins to heat. The laws of thermodynamics dictate that the earth, once heated, must *radiate* energy back toward space. Because the earth is much cooler than the sun, this occurs in much longer wavelengths than those of incoming solar radiation: specifically, as far infrared radiation in a band of 3–30 μm, with most in the 8–15 μm band. We don't see infrared energy; snakes do, but we don't. We can, however, sense it as the radiant heat coming from a warm rock on a cold day.

Water vapor varies from virtually 0 to 4 percent of the total volume of the atmosphere and efficiently absorbs infrared radiation in 5–8 μm wavelengths and in wavelengths exceeding 13 μm. Carbon dioxide accounts for an average of only 0.03 percent of the atmosphere, but it avidly absorbs infrared radiation at 4 and 13–17 μm. Consequently, water vapor and carbon dioxide—both poor absorbers of *incoming* solar radiation— are very successful at collecting *outgoing* infrared radiation. On clear days, outgoing radiation in wavelengths between 8 and 13 μm tends to escape into space. Clouds (not just invisible water vapor, but clouds) are very effective at absorbing all outgoing radiation but especially in the 8–13 μm

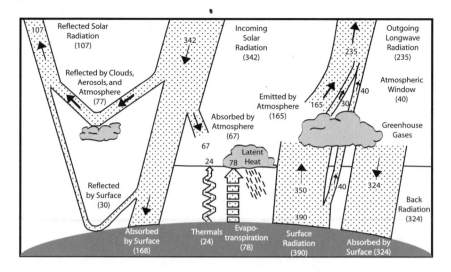

A schematic diagram showing the earth's energy balance. All values are in watts per square meter. (Based on Kiehl and Trenberth 1997)

window. Who hasn't noticed that, everything else being equal, a cloudy night will always be warmer than a clear night?

The atmosphere is thus heated more by outgoing radiation from the earth than directly by the sun. We intuit this when we observe that the hottest part of a summer day is not at noon, but a couple of hours later. Heat is not just stored in clouds; it is also reradiated back down to the earth. If the atmosphere did not absorb outgoing radiation, the earth's surface would average −23°C instead of +15°C. There would be no free water and probably no life.

The atmosphere also gains heat from the earth through two additional processes beyond reabsorption of outbound radiated heat. Air is directly warmed by *conduction* within a few centimeters of the earth's surface. As it heats, the air expands and grows less dense. As it becomes more buoyant, it tends to rise, thus transferring heat into the upper atmosphere by *convection*. Heat can also be transferred from water into the atmosphere as *latent heat*. Heat must be added to water as it jumps from a low-energy liquid to a higher-energy gas, a process called *evaporation*. Heat is transferred into the atmosphere when that water vapor condenses, leaving heat behind as the liquid water falls.

Direct transfer of energy from the surface contributes slightly less than 5 percent of the atmosphere's warmth; latent heat of evaporation adds another 15 percent. Absorbed radiation—either directly from the sun (13 percent) or radiated back from the earth's surface (67 percent)—accounts for the rest of the atmosphere's heat. Coupled mechanisms thus transfer heat from the sun to the earth, mix it upward into the atmosphere, and then lose it in space. What happens now?

Let's retreat to that simplified model we've been building. Assume that the earth is a homogenous sphere bathed in parallel rays radiating from an infinitely distant source. Rays at the equator arrive perpendicularly, delivering their full 1.94 calories to each square centimeter each minute. Rays obliquely striking the Arctic and Antarctic poles, however, glance across the surface. The polar ice caps, with their high albedo, succeed in deflecting most of what little energy might have been absorbed.[4] This geometry sets up a tremendous inequity of heat distribution on the earth's surface: hot equator, cold poles, and an uneven gradient in between. Entire sciences—meteorology, climatology, atmospheric physics, oceanography—revolve around this gradient.

In 1735 George Hadley surmised that great cells of circulating air rise at the hot equator, flow poleward high above the Northern and Southern Hemispheres, and then descend on their respective poles. These *Hadley cells* were suggested by observations of persistent low atmospheric pressure at the equator and high pressure at the poles. Hadley cells alone, however, do not address a few complicating factors, such as the earth's rotation and the tendency of rising air to locally cool and lose its moisture as tropical rainfall. According to Hadley's model, winds near the ground at midlatitudes in both hemispheres should complete this loop, consistently blowing back from the poles toward the equator. Such is not the case, however. The real world has more convoluted patterns of pressure and wind than those predicted by this single-cell model.

Hadley was on the right track, but his model needed refinement. Instead of a single cell on each side of the equator, it turns out that a three-cell model in each hemisphere better approximates the real world. Hot, moisture-laden air does indeed rise from the Intertropical Convergence Zone (ITCZ) near the equator. The ITCZ is thought of as a convergence

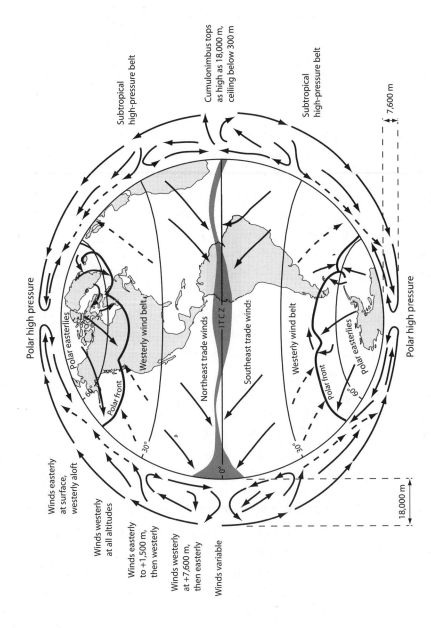

Subtropical
high-pressure belt

Cumulonimbus tops
as high as 18,000 m,
ceiling below 300 m

Subtropical
high-pressure belt

7,600 m

Polar high pressure

Polar easterlies

Westerly wind belt

Polar front

Northeast trade winds

I.T.C.Z.

Southeast trade winds

Westerly wind belt

Polar front

Polar easterlies

Polar high pressure

30°

0°

30°

60°

60°

Winds easterly
at surface,
westerly aloft

Winds westerly
at all altitudes

Winds easterly
to +1,500 m,
then westerly

Winds westerly
at +7,600 m,
then easterly

Winds variable

18,000 m

A three-cell model of the earth's atmospheric circulation. (Based on Pinet 1998) zone because surface winds converge into it from both the north and the south to fill the void created by its rising air. Underlying the ITCZ is much of the world's Tropics, wetted perpetually by rainfall generated in that rising moist air.

The ITCZ migrates north and south a few degrees above and below the equator in response to summertime heating first in one hemisphere, then in the other. The rising air cools with altitude and dries when its moisture and heat are lost as tropical precipitation. Rather than continuing to flow farther from the equator, much of the cool dense air descends back toward the ground at about 30° north and south latitude. This descending air piles up at the surface, creating the stationary high-pressure zones that girdle the globe in two rings of bright skies and infrequent precipitation. This is exactly where we find the world's great deserts: the Sahara, Atacama, Great Australian, and North American Deserts.

A pair of low-pressure zones are formed at 60° north and south latitude. Here surface winds moving away from the midlatitude deserts collide with cold air that has descended onto and then away from the poles. This collision can be violent. The stationary Aleutian Low, centered just below 60°N off the Alaskan coast, is a breeding ground for the great winter storms that rake North America. The analogous stationary low in the Southern Hemisphere lies entirely over the ocean basins north of Antarctica, with no landmass to block or dampen its storms. Consequently, these are among the wildest waters in the world, the "furious fifties" of Drakes Passage south of the Atlantic Ocean.

Cool, dry air descending to the surface must diverge when it hits the earth, initially flowing either north or south away from a zone of high pressure. Conversely, when hot, moisture-laden air ascends, an area of low pressure is created, and surface-level winds flow in from the north and south. This area tends to form in a band on or parallel to the equator.

With these diverging or converging surface winds, we run into another complication. The earth spins from west to east on its axis. This spin causes a point on the equator to rotate east almost 1,600 km each hour; a second point at 60° latitude is traveling at only half that speed. An object launched from the equator toward Greenland will still have the equator's 1,600 km/hr eastward velocity as it flies north. If its northward path is uncorrected, the object will land somewhere in Russia rather than in Greenland. This is called the *Coriolis effect*, which dictates that all objects moving north or south in the Northern Hemisphere apparently tend to veer to the right, while those moving in the Southern Hemisphere veer to the left.

Air masses flowing away from zones of divergence (or toward zones of convergence) will be steadily deflected either right or left, depending on the hemisphere in which they are located. Tropical winds in the Northern Hemisphere that converge upon the equator are deflected to the right, that is, toward the west. The same winds in the Southern Hemisphere are deflected to the left, which is also to the west.

We label wind by the direction whence it comes, so that a wind blowing from the west is called a *westerly*, and a wind from the east is an *easterly*. Perversely, ocean currents are named the opposite way: westward-heading currents are called *westerly currents* and vice versa.

A three-cell model of the atmosphere, integrating the Coriolis effect, consists of westerly winds dominating the midlatitudes and easterly winds converging from the Tropics toward the equator. The winds between 30° and 60° north and south latitude are known as the *extratropical westerlies*. Much of the world's temperate climatic zones underlie these westerlies. Easterly winds converging from the Tropics toward the equator are known as the *trade winds*, primarily because ships sailing between the Old and New Worlds depended on these winds for westward travel. This model explains observations of wind in the real world more accurately than Hadley's original single-cell model.

Another geometric phenomenon creates the seasons of the year. The earth spins on an axis tilted 23° away from the plane of its revolution around the sun. In its yearlong tour around the sun, the earth with its two hemispheres is subject to inverse periods of higher or lower amounts of solar input—summer or winter—depending on which hemisphere is pointing toward or away from the sun. On the short days of winter, less solar radiation arrives in that hemisphere to be absorbed and heat the ground and atmosphere.

Temperature gradients drive the weather phenomena that affect the earth. In the Northern Hemisphere, this equator-to-pole temperature difference varies from 43 °C in winter to only 10 °C in summer. Winter-time weather tends to involve clashes between more disparate bodies of warm and cold air. Winter storms therefore tend to grow bigger and affect larger areas than summer storms. Many summer storms tend to form from air rising from localized hot land or warm water.

Let's investigate another fundamental complication of global atmospheric circulation. The surface of the earth is not homogeneous; it is composed of oceans and continental landmasses. Water and land behave very differently when heated. Water at a given temperature has the capacity to hold five times as much heat as a similar volume of granite. Not only are the *heat capacities* of these substances different, but the rates at which they absorb heat also vary by a factor of two: soils heat and cool more than twice as fast as water when similar amounts of energy are supplied. The oceans, though experiencing lower-amplitude temperature swings, hold much more heat than the land surfaces.

Thirty percent of the earth's surface is covered by land. Here air temperatures climb higher and more quickly than over water; thus inland temperature ranges will always be greater than those along a coastline. The huge landmass of Asia is subject to the world's greatest temperature gradients. At Yakutsk in Siberia, the average seasonal cycle of temperatures varies by 62 °C, a range never found over open water. This continental heat can produce transitory zones of low pressure. Unlike the lows over equatorial oceans, land-derived lows are usually relatively devoid of water (although the Amazon Basin is a glaring exception). Continental effects are more pronounced in the Northern Hemisphere, because the earth's landmasses are predominantly bunched up north of the equator.

The flow of energy from the hot equator to the cold poles is not a smooth process; it seeks but never quite finds an equilibrium. If this flow were uniform, we would never experience storms. In the real world, heat builds up in the oceans and is released in unstable bursts, such as hurricanes churning through the equatorial oceans or winter storms blowing onshore from the northern oceans.

As water in the subtropical ocean warms, it approaches a point at which it increasingly prefers to exist as a vapor rather than a liquid. We think of the boiling point of water as being fixed at 100 °C, but it actually depends on the vapor pressure of the water (determined by its temperature and salinity) and the vapor pressure of the overlying air (determined by its pressure, temperature, and existing moisture level). Evaporation occurs below 100°C, but it's a two-way street: Water molecules zip back and forth across the air-water boundary, maintaining a dynamic equilibrium between the two states. As the system temperature climbs toward 100°C, more and more molecules choose to fly rather than swim.

In reality, it's very difficult for seawater to warm above 30°C. Beyond this temperature, net evaporation accelerates, and the overlying air begins to get stirred up. This brings in drier air that allows evaporation to occur even faster, effectively cooling the water. Evaporation of each cubic centimeter of water removes 550 calories from the ocean, as seawater jumps from a liquid to a gaseous state. This latent heat is carried with the water vapor into the atmosphere. Summer thunderstorms building over Florida feed on this heat and moisture. Tropical cyclones in the western Pacific metamorphose into unstable beasts as they incorporate this heat.

The calories do add up. The energy within an average thunderstorm is equal to that released by the atomic bomb that destroyed Nagasaki in World War II. A tropical hurricane is driven by ten thousand times as much energy. Hurricane Hugo, for example, had twelve days to fester over the warm waters of the Atlantic Ocean in September 1989 before it exploded against South Carolina with winds of 220 km/hr. The hurricane hurled full-blown tornadoes at North Carolina: arrows thrown angrily by a rampaging god. A 6 m tidal surge smashed against Awendaw, South Carolina. Had half a million people not evacuated their homes along the coast, fatalities would have been much higher than the fifty-seven that did occur on the U.S. mainland.

Air that moves steadily, such as the midlatitude westerlies or the tropical trade winds, pushes on water at the ocean's surface and creates saltwater currents. Like air currents, these ocean currents are susceptible to the twisting contortions of the Coriolis effect. Ocean currents are also deflected by continental margins. These two factors combine to establish five great circulating *gyres* that dominate flow within the Indian, Atlantic, and Pacific Oceans. Moving water can pile up against the edge of a continent or at the point where two currents collide, producing a slight but

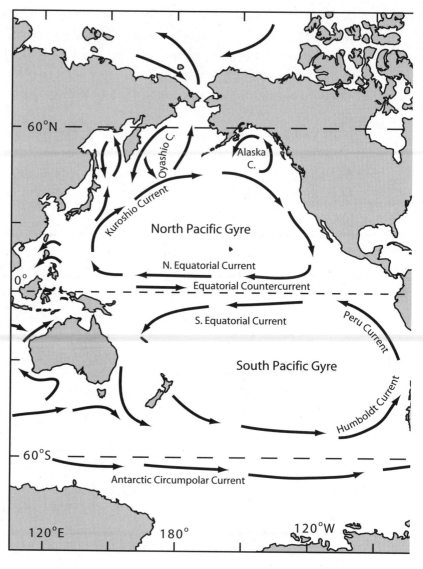

Global patterns of oceanic surface-water currents. (Based on Pinet 1998)

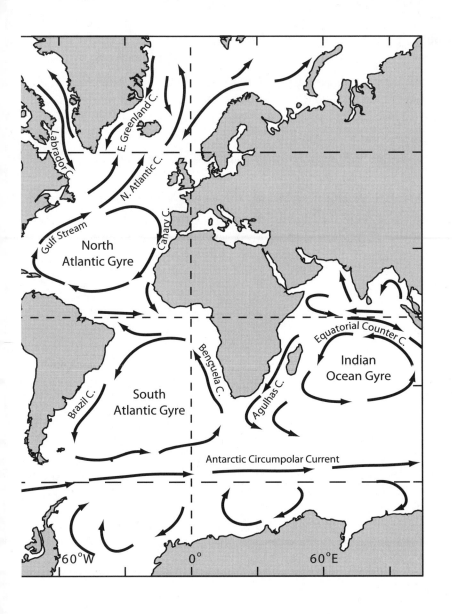

very real bulge in the ocean's surface. Even though this bulge is rarely more than 1 m in height, it has the potential to slide back downhill when forces against it relax. Subtle salinity differences (especially when coupled with temperature gradients) change the density of ocean water sufficiently to create currents that sweep through the Atlantic with the volume of a hundred rivers the size of the Amazon.

The earth is a weather machine with a complicated array of humming components, each interlocked with the other. Solar energy, passed through the atmosphere and absorbed by the earth's landmasses and oceans, powers this system. The minute-to-minute redistribution of this energy is what we call *weather*; the long-term patterns of redistribution are what we call *climate*.

The Geology of Climate

Climate is the longer-term atmospheric expression of many different variables, some changing quickly, some gradually. Geologic factors, such as the arrangement of continents or the connectedness of oceans, change so slowly that they might be considered less as a confusing variable and more as the framework within which climate and climate change occur. The rules dictated by geology do evolve, but only very gradually.

Tectonics

The theory of plate tectonics explains the migration of continents around the earth. With sufficient imagination, it's possible to play "movies" of the continents as they crash together or drift apart. A ten-minute film spanning the 250 million years since the Permian Period (with a frame shot every 10,000 years) would start with all of the earth's continents clustered together in a single landmass known as Pangaea. Within the first minute, North America and Asia are sailing north from Africa. Then Africa and South America split away from Antarctica. After five minutes, as the continents spread around the globe, the Atlantic Ocean begins to widen. India collides with Asia as Australia finally breaks away from Antarctica. The Pacific Basin widens, and only in the last minute does Central America rise to connect North and South America.[1] Not surprisingly, worldwide climate regimes shifted dramatically as this through-flowing equatorial passage closed.

Heat flow around the globe can radically change as oceans rearrange themselves. Twenty million years ago, South America sailed away from Antarctica. As Drakes Passage between the two opened, the Southern Ocean began to whirl without interruption from west to east between latitudes 55°S and 62°S. Suddenly Antarctica was oceanographically and then atmospherically isolated from the buffering effect of the rest of the world. It nose-dived into a frigid environment that locked vast amounts of water into glaciers thousands of meters thick. With the Southern Ocean isolating Antarctica from the southern tips of Africa, Australia, and the Americas, the stage was set for a more equable present-day climate in the midlatitudes of the Southern Hemisphere. Seasonal cycles Down Under are much less pronounced than in the Northern Hemisphere.

Three million years ago, the landmasses of North and South America became connected at the Isthmus of Panama. Ocean currents, driven by trade winds, had previously been able to flow without interruption from the Atlantic into the Pacific. After the isthmus rose, these currents were deflected north and south, spiriting tropical heat away from the equator. If the Atlantic and Pacific Oceans could freely mix today, our climate would be radically different. England would be an icebox.

Another climatological implication of plate tectonics involves the clustering of great landmasses. Continental masses, especially those at or near either pole, lose heat much more quickly than oceans during winter. Many of the world's continental landmasses had congregated around the South Pole 250 million years ago. All of Antarctica and portions of South America, Africa, and India were then buried beneath glaciers larger than any that exist today. As a result, much of the world's free water was locked up as ice during the Permian Period, and sea levels were lowered by many tens of meters. Pangaea, as that agglomeration of continents was known, has since broken into the continents that we recognize today.

Currently Antarctica holds 90 percent of the world's fresh water in its 29.3 million km^3 of glaciers.[2] By comparison, the North Pole's ice cap is vanishingly thin, containing less than 1 percent of that found at the South

Glacial ice descending thousands of meters from peaks on Anvers Island into Gerlache Strait, Antarctica, 1999.

Ice off Enterprise Island on the Antarctic Peninsula, 1999.

Pole. Unlike Antarctica in the Southern Hemisphere, which is isolated by its encircling Southern Ocean and uninterrupted Antarctic Circumpolar Current, the continents in the Northern Hemisphere divert water (and heat) toward the pole. A thin layer of ice floats on top of the Arctic Ocean, warmed by seawater pumped up from the midlatitudes.

Mountain Building

Geologic processes can influence global climate in other ways. As mountains rise, rocks are exposed to weathering processes, releasing silica and calcium. Carbon dioxide–rich groundwater percolates through the rock, recombining ions to form calcium carbonate and silicon dioxide. The end result is that CO_2 is extracted from the atmosphere and dumped on the ocean floor. With the loss of CO_2, the entire atmosphere cools.[3]

The Himalaya Mountains and the Tibetan Plateau began to rise after India crashed into Asia 50 million years ago. Uplift occurred in a spasmodic fashion in response to both the initial compression and crustal adjustments that followed the collision. A rapid phase of uplift occurred about 7 million years ago. Carbonate rocks deposited throughout the world at that time document a precipitous decline in atmospheric CO_2. Simultaneously, Asian monsoons suddenly strengthened.[4] Uplift, atmospheric response, and climate change: They are all related over the long term.

High mountains such as the Himalayas also influence the atmosphere in another manner. Rocks 7,000–8,000 m above sea level may seem cold, but like rocks anywhere else, they absorb incoming solar radiation and then radiate heat back into the atmosphere. The rock-air interface at these 200-millibar (mb) altitudes injects heat directly into a rarified atmosphere. A unit of energy goes a lot farther toward raising air temperature here than in the thicker air found at sea level. This heating strongly influences wind patterns that are important determinants of monsoon vim and vigor.

Half a world away, California's Death Valley illustrates yet another climatic response to geologic processes. This desert basin lies in the shadow of the Sierra Nevada. These mountains did not begin rising until 3 or 4 million years ago: relatively recently in geologic terms. They continue to rocket 3 or 4 cm skyward every hundred years.[5] The Sierra Nevada trap much of the moisture that blows in from the Pacific. Death Valley, hiding in the rain shadow of the mountains, limps along with less than 50 mm of rain a year.

Volcanism

Crustal plate arrangements, ice caps, and mountain building constitute the grand geologic arena within which the climate game is played. Volcanism is another geologic process that influences climate, but its effects can be either short or long lived. Flood basalts periodically (perhaps every few tens of millions of years) erupt at hot spots around the earth. The Deccan Traps were formed when southwestern India was buried beneath 2 million km^3 of magma.[6] Most of this material came to the surface within a single million-year period, delivering so much heat so quickly from the earth's interior that it was probably a major factor in the climate change that occurred at the end of the Mesozoic Era, about 65 million years ago. This set the stage for the even more dramatic impact of a meteorite, which

An eruption of Mount Pinatubo in the Philippines on June 15, 1991. (Photograph by E.W. Wolfe, courtesy of the U.S. Geological Survey)

was probably responsible for the mass extinction of dinosaurs and other species around the world at about the same time.[7]

Eruptions of individual volcanoes have temporarily forced change in the atmosphere.[8] Sometimes the changes are merely cosmetic. The 1991 explosion of Mount Pinatubo in the Philippines caused little more than spectacular sunsets around the world for a few months. At other times, however, the changes have had significant meteorologic implications. Benjamin Franklin conjectured that the particularly cold winter of 1783–84 was due to sunlight being reflected back into space after a large Icelandic eruption. Three decades later, Tambora Volcano on the Indonesian island of Sumbawa exploded, ejecting 150 km³ of pumice and ash into the atmosphere: one hundred times more than Mount St. Helens blasted into the skies above the state of Washington in 1980.[9] A year later in 1816, North America and Europe both experienced the "year without summer," when crops failed and snow fell in June.[10]

The 1982 eruption of El Chichón in Mexico threw huge quantities of sulfur-rich gases into the atmosphere that took more than a year to settle back to earth. Sulfurous gases and fine particulates were blasted into the stratosphere; sulfuric acid droplets both absorb incoming solar radiation and reflect it back into space. This may have been partially responsible for the estimated 0.3°C global cooling measured during the two years after the El Chichón eruption.

For the most part, though, these atmospheric effects were transitory, lasting only as long as the volcanic aerosols remained airborne. Scientists have explored the possibility that large submarine lava flows (at least 10 km³ in size) may have occasionally heated ocean water enough to upset the world's climate patterns.[11] Volcanic dust can either raise or lower sea surface temperatures, depending on whether it circulates in the atmosphere at low or high latitudes.[12] If aerosols are present only in the Tropics, sea surface temperatures can rise.[13] Can short-term perturbations, such as volcanic dust, trigger atmospheric instabilities that would produce longer-lasting climatic change? At this point, we don't know, though many scientists would argue that the answer is "no." It might just depend on the scale of the eruption.

El Chichón erupted in 1982, just before the greatest El Niño of the twentieth century. Was the subsequent worldwide atmospheric temperature drop causally related or merely coincidental? Without the El Chichón eruption, would the 1982–83 El Niño have been an even warmer event? The summer of 1816 was remarkably cold, but it was within the bounds of known climate variability; again, was this caused by a volcanic eruption, or was it mere coincidence?[14] Scientists at the University of Washington calculated that large eruptions such as Tambora decrease worldwide air and sea surface temperatures for a year or two. They could, however, find no paleoclimatic evidence that such volcanic explosions had a demonstrable effect on the worldwide precipitation patterns or sea level pressures that are integral components of El Niño events.[15] Because climate change is so fundamentally tied to energy transport, it is tempting to try to relate it to volcanic activity.[16] For the most part, however, this has remained an elusive connection.

Geomorphology

The whims of weather can affect individual basins or entire continents. Geomorphic features, such as the shape, placement, and size of river basins, can be lenses that locally bring climate into sharp focus. Scientists have noted that spring runoff has arrived increasingly early in recent decades throughout a series of basins with headwaters in the Sierra Nevada.[17] Weather stations in California have shown a trend toward warmer winters since the early 1940s. Not surprisingly, basins whose primary catchment areas were at moderate altitudes were most sensitive to this change because snow there is closer to its melting point. This shift in runoff timing is important because it connects floods across basins and allows them to be examined in the light of regional (or larger) climate changes.

The larger patterns of climate affect broad regions, not just individual basins. It's true that each basin in a geographic area will respond uniquely to precipitation, depending on its elevation, preexisting saturation, and channel condition. It is possible and even necessary to look beyond a given basin if one is to arrive at a meaningful understanding of river response to climate change and the likelihood of flooding events.[18] Streamflow can be viewed as the integration of precipitation.[19] The correlation between these phenomena is stronger in some regions than in others. For instance, rainfall and streamflow both rise in the American Southwest and southern Alaska (and fall in the American Northwest) when specific patterns of atmospheric pressure and sea surface temperature exist over the equatorial Pacific Ocean.[20]

The geomorphic consequences of climate, precipitation, and streamflow can be dramatic, even tragic. Four million cubic meters of rock and mud broke loose from a mountainside and slid into Spanish Fork Canyon near Thistle, Utah, during the torrentially wet spring of 1983. Thistle was drowned in the lake that formed behind the landslide. Three highways and a railroad lay beneath the 5 km lake. Four hundred million dollars later, the highways and railroad were reopened.[21] The town is just a memory now.

The Thistle landslide in Spanish Fork Canyon in Utah. This landslide was re-activated during intense winter rains in the Wasatch Range associated with the 1982–83 El Niño.

Geologic processes, such as volcanism, mountain building, and plate tectonics, set the large stage on which climate acts. Specific geomorphic features, such as a drainage basin's shape, size, and altitude, can shape the regional expression of climate on a smaller scale. Whether taken on a large or small scale, geology does impact climate.

Winds of 5 Change

Most of us, living in cities and working indoors, can remember only a handful of major weather-related events. I recall humming Christmas carols while struggling through the deep snows that blanketed northern Arizona in 1967. My kayak was out of control on the normally bone-dry Rillito Wash, careening past Tucson on the 841 m³/s flood of October 1983, the largest in anyone's memory. Parched ponderosa pines ticked in the heat as I nervously scanned Flagstaff's western horizon for signs of fire during a drought in June 1996. We remember unusual atmospheric events not for their own sake, but because of their specific impact upon our own lives.

Beyond the spectacular exceptions, most events fade quickly from memory, diluted by the immediacy of each day's sunshine or rain. With so few personal data points, it can take years for an individual to construct a long-term perspective on climate. Over a lifetime, many people eventually develop a sense of what is "normal." In the mountains of northern Arizona, it's usually safe to plant beans by mid-May; summer thunderstorms ought to start in mid-July; the fall display of aspen colors should peak after the first week in October.

Climate is a pattern of meteorologic events observed over a defined period of time. That period is determined by how far back we are able— or willing—to see into the past. Extend or shorten the period, and observations about "average" climate will inevitably change. Data from a local weather station typically span a few decades, perhaps back to the Second World War or some other paroxysm that spawned a new airport or another town along a railroad. Sporadic climatological notes can be gleaned by reviewing historical accounts spanning a few hundred years. Beyond written history, earth scientists are beginning to interpret fossil ice, corals, and tree growth, which provide proxy records of climate stretching thousands or even hundreds of thousands of years into the past.

When viewed from a vantage point of years rather than weeks, movement of the atmosphere is best described as climate: an historical perspective that incorporates not only the long-term average of weather, but also those excursions from the norm that are part of the average. Extreme weather, with its floods and droughts, is part of the entire package that we call *climate*. Bear in mind, though, that climate itself can vary, and any given flood or drought may turn out to be the leading edge of a new climate regime. It's difficult but worthwhile to distinguish the "noise" of weather from the more meaningful and long-lasting cadences of climate.[1]

Climate is what you expect; weather is what you get. It's all a matter of differing scales of time. As Kelly Redmond, a climatologist with the Desert Research Institute in Reno, Nevada, remarked, "Climate to us is weather to a glacier." The single most important idea that emerges is that *climate patterns can and will vary with time*. In other words, the rules will change even as we play this game.

How fast can climate vary? What natural (or human) pressures initiate such change? Do multiple pressures exert their influences simultaneously? Many processes can bring about climate variation. Each operates a little differently and proceeds at its own pace. Some processes are more pronounced over oceans than over land. Some exert more pressure at the equator; others are more effective closer to the poles. Some changes wax or wane over centuries; others vary over just a few years. The task of teasing out even the two or three most dominant processes can be daunting. Perhaps at this point, it is enough to understand that multiple processes do occur simultaneously, each contributing to the weather and climate within which we live.

One of the most intriguing lessons learned from the study of climate over the last half century is that interactions between oceans and the atmosphere appear to be driven by an internal clock, ticktocking through episodes of warm and cold climate extremes on scales of years to decades. Left to its own devices, the earth's climate would oscillate between states of high and low energy. But this internal rhythm is modified by external forces: fluctuating amounts of energy delivered by the sun and released from within the earth. These external variables accelerate or decelerate the innate oscillations that would otherwise be governed solely by the oceans and atmosphere. In the largest sense, floods and droughts are influenced by both internal and external forces. On a human scale, however, internal oscillations of climate are probably more germane. Nevertheless, let's first look at the external variables; subsequently we will dive into the internal aspects of climate variability.

External Atmospheric Variables

Geologists at the turn of the last century were fascinated by irrefutable evidence of the Ice Age, that time during the Pleistocene when ice sheets intermittently enveloped much of the northern world from 1.6 million to about 10,000 years ago. Rocks throughout the high latitudes of North America were gouged by glaciers that scratched southward from the pole, scooping out the Great Lakes as they advanced. Vegetation patterns were shifted dramatically downslope and toward the equator by a climate that was generally much colder and wetter than that of the previous period or even today. Alpine conifers grew then where grasslands are found now. As geologists dug further into this subject, they realized that the Pleistocene Ice Age had been not one but a series of glacial advances, an oscillating series of climate fluctuations. The shift in climate patterns during the Ice Age was of such a magnitude that it represented a true change more than just a variation.

Milutin Milankovitch was a Yugoslavian meteorologist who in 1920 examined variations in the earth's rotation about its axis and in its orbit around the sun. The earth has an elliptical orbit, its axis of rotation is tipped away from the sun, and it wobbles as it rotates. Milankovitch worked out the cyclic variation in each of these three phenomena, with periodicities of 92,000, 40,000, and 21,000 years, respectively. Solar radiation received by the earth slightly increases or decreases as these cycles wax and wane. At times, the warming or cooling effects of one cancel out another, but over longer intervals, their effects align and are magnified.

Could the individual advances and retreats of glaciers during the Ice Age be explained celestially? Glaciologists have described seventeen (more or less, depending on the continent on which one stands) major cold episodes during the Pleistocene: an average of one every 100,000 years. This interval is temptingly close to the longest of Milankovitch's three celestial cycles. Marine geologists used oxygen isotopes to determine temperatures of deep sea sediments at the time of their deposition in the Caribbean Sea and equatorial Atlantic Ocean.[2] They found evidence for cold peaks every 40,000 years or so, again resonating with the cycles calculated by Milankovitch.

But deep sea muds do not an ice age make. The celestial mechanics that determine variations in the solar radiation received by the earth must have been operating for billions of years. Glaciers came and went during the Permian Period, 250 million years before the Pleistocene. Milankovitch cycles can not explain the intervening hiatus. In the Permian and again during the Pleistocene, other variables must also have been operating to lower the earth's threshold for descent into an ice age.

One of those variables may be evident in the highlands of central Asia. China's Loess Plateau is covered by a thick, finely striated blanket of dust. This was deposited by the East Asian monsoon system, driven by thermal differences between the Asian landmass and the Pacific Ocean. Winter brought dust with the dry cold weather, while summer was marked by wet warm rain and a type of soil development that may have extended as far north as the Mongolian border. The monsoon system reflected in the dust shows strong periodicities reverberating at 100,000-, 41,000-, and 19,000–23,000-year intervals. The Tibetan Plateau experienced two geologically driven growth spurts, one at 7.2 million years ago and another at 3.4 million years ago. The first could have initiated or strengthened this East Asian monsoon system; the second may have been instrumental in triggering the Pleistocene Ice Age. Dust particle sizes in the Loess Plateau changed around 2.6 million years ago, a transition that marks an enhancement of winds in response to growth of the polar ice cap.[3]

Cycles are predictable. They offer the reassurance that our world is an orderly place after all. But close examination of paleoclimate data fails to reveal a clean periodicity in temperature, moisture, or other determinants of climate. We are left with the less satisfying observation that the flow of energy into and out of the earth and its atmosphere is not always balanced. Such imbalances are responsible for long-term climate change. Are there other scales of change in the climate record? The answer is "yes": both longer- and shorter-term scales are evident. In chapter 4, we examined prolonged geologic processes that shape the atmosphere, such as the migration of continents, the tectonic sculpting of major drainage basins, and the production of geothermal energy. Now let's consider other factors that may spur shorter-term variations in climate.

Asian astronomers commented on sunspots as early as A.D. 1077. Sunspots are dark blotches on the sun's surface with temperatures 1,400°C cooler than surrounding areas. Galileo Galilei studied them through his newly invented telescope in 1611. Heinrich Schwabe, a German pharmacist, plotted the number of sunspots each year in the early 1800s, noting a regular periodicity of about ten years. Rudolf Wolf reviewed Schwabe's notes and the previous 150 years' records and refined this period to 11.1 years.[4] The number of sunspots typically varies from as few as five to as many as one hundred. Solar radiation varies with the number of sunspots: The more sunspots, the more energy is released into space.

Sunspot activity has fluctuated widely during the past millennium. Between the years A.D. 1100 and 1300, an unusually high number of spots were recorded. Conversely, fewer sunspots occurred during the seventy years from 1645 to 1715 than normally occur in a single year. This period generally corresponds to the Maunder Minimum, a time of exceptionally low temperatures in Europe during the Little Ice Age that stretched from the late 1300s until 1850. Modern measurements of the sun's brightness and temperature have demonstrated short-term changes: an 11°C drop at its surface in 1977; a 0.07 percent drop in solar radiance measured by satellites outside the earth's atmosphere from 1981 to 1984. These changes can and do happen. The relevant question is this: Are they big enough and do they persist long enough to affect our climate?

Computer modeling suggests that a 2 percent decrease in solar radiation lasting fifty years could initiate glaciation. Tree-ring records hint at a tantalizing association between vegetation growth and sunspot activity. Sea surface temperatures, atmospheric pressure gradients, and rainfall have all been noted to swerve in response to solar activity.[5] The connection between solar activity (i.e., sunspots, magnetic field variation, and solar wind) and its effects on the earth's surface is disconcertingly complex. Changes in incoming ultraviolet radiation may affect the thermal structure of the stratosphere; solar winds may induce high cirrus cloud formation in the troposphere; cosmic rays may lead to nitric oxide formation, which causes depletion of ozone and cooling in the stratosphere.[6] But no clear signal has yet been found that tightly orchestrates solar cycles with the entire symphony of climatic phenomena.

In all likelihood, the earth is subtly influenced by orbital and radiation changes, but the atmosphere responds complexly with multiple interconnected feedback loops. It makes sense that variations in solar radiation and the earth's orbit should affect our climate, but sometimes it's just too great a stretch between *should* and *how*. In other words, sunspots are unreliable predictors of weather, and by themselves orbital cycles can not foretell climate. The rhythms of climate variability are more likely to be sensed from within the air/ocean system than from outside forces.

Internal Atmospheric Variables

Milankovitch cycles and sunspots are external forces that exert small but sometimes significant pressures on the earth's climate. There are internal forces as well, which are best exemplified by the much ballyhooed phenomenon called *El Niño*. The earth's climate is a fundamentally dynamic system capable of huge and significant internal fluctuations and oscillations. These internal oscillations "ring" in complicated harmonies as different components of the oceans and atmosphere interact over periods ranging from years to decades. Taken together, these oscillations are among the most fundamental aspects of climate. Tiny external forces may tip the internal balance just enough to drive climatic probability one way rather than another for awhile. But it is the internal rhythm of climate that creates the recurring theme song of flood and drought.

The world's atmosphere has heated by 1–2°C for centuries at a time in the not-so-distant past. The Grosser Aletsch Glacier, which is characteristic of many ice fields in the Swiss Alps, regressed during the Medieval Warm Period from A.D. 900 until about 1300.[7] During this time, Erik the Red sailed to the west coast of southern Greenland, where he found grassy valleys and great fishing (at least, that's what his brochures claimed). Seal meat was a central staple of the colonists' diet; they also shepherded caribou and cattle and probably planted cereals.[8] No doubt Greenland offered a precarious existence, with life clinging to this thinnest edge of survivable conditions. Times were as good as they get, and the climate was benign for a century or two after the colonies were founded. The western settlement in what is now the Godthab District had ninety farms and four churches. The environment, however, changed drastically between 1341 and 1362. The thin Greenland edge could no longer support the Norsemen, and they abandoned their colonies. The Little Ice Age had begun.

Thousands of miles away in the American Southwest, people were slowly populating the deserts and canyons of the Colorado Plateau. These people, once called the Anasazi, have now been burdened with the clumsy if politically correct name of Ancestral Puebloans. Early members of this group had lived in simple pit houses since A.D. 500. This culture reached its apex after about 900, when communities in and around Chaco Canyon exploded onto the scene. Irrigation systems blossomed across the desert, and roads stretched hundreds of miles out from Chaco Canyon. The canyon settlement itself fell into decline after 1132, but other communities, such as those at Mesa Verde and Kayenta, flourished until almost 1300. Then suddenly, the Ancestral Puebloans departed from their plateau homeland.

For centuries these people had proven remarkably adaptable to the desert's varying demands, but something snapped. What drove them away? Many explanations have been advanced: drought, disease, or internal strife. Perhaps it was foreknowledge of the impossibly clumsy name with which they would eventually be saddled. Jeffrey Dean, studying tree-ring series throughout the Southwest, has suggested another fascinating possibility.[9] The Ancestral Puebloans flourished during a prolonged period of relative climatic stability that began around A.D. 1000. It's true that the region was disturbed by drought from 1130 until 1180 and again from 1275 until 1300. By and large, however, the Ancestral Puebloans were blessed with three centuries of unusually stable weather patterns. Even though they probably lacked those little seed packets impaled on sticks at the ends of each garden row, the Ancestral Puebloans could plant their corn, beans, and squash with a comfortable degree of certainty about what would come up and when. Dean's most interesting conjecture is that, from 1250 through 1450, the Colorado Plateau was subject to a wildly varying climate; very wet years immediately followed very dry years. The increased unpredictability may have been the last straw for the Ancestral Puebloans.

The Norse colonies in Greenland and the Ancestral Puebloan expansion in the Southwest both flourished during a centuries-long climatic optimum. Climates can also vary decade by decade. Americans have not yet forgotten the Dust Bowl days that began in mid-1933: a severe drought that lasted three years. By 1935, 80 percent of the Great Plains had suffered some degree of wind erosion. But compare our Dust Bowl experience with that of modern-day sub-Saharan Africa.

The Sahel Desert (which includes parts of Senegal, Mauritania, Mali, Burkina Faso, Niger, Nigeria, Cameroon, and Chad) continues to be gripped by a drought that began forty years ago. Malam Garba, a farmer in Dalli, Niger, remembers villagers hunting antelope, monkey, wolf, fox, squirrel, and rabbit before the drought. Now there is no game. Forty years ago, he and his brother grew enough millet to feed their families and provide a surplus for sale; now they cultivate three times as much land but harvest only one-seventh as much grain.[10] Sand dunes lap at their windows as dry wind strips away their soil and hope.

From north to south, the Sahel region of western Africa was anywhere from 37 to 15 percent wetter than the long-term mean throughout the good years in the 1950s. Crops flourished then. Beginning in the early 1960s and persisting through the present, rainfall abruptly decreased: 31

to 13 percent below the long-term mean in the 1970s, and 24 to 20 percent below normal during the 1980s.[11] The consequences have been dire. This region, verdant enough to export grain at midcentury, is now haunted by famine, disease, and the dislocation of untold thousands of inhabitants. To make matters worse, the Sahel's population has risen from 60 to 107 million during this drought.[12]

What mechanisms could produce this change? First and most obvious is the clearing of 123 million hectares of natural vegetation for agriculture.[13] Native perennial vegetation was replaced by crops such as millet, peanuts, and sorghum. Once a farmer depletes his field of nutrients, he simply abandons it and plows up new land. As trees are removed, the wind moves at ground level like a scythe, grasping at the vulnerable top soil. The director of Niger's National Department of the Environment estimates that an area the size of Luxembourg is lost each year in his country to this degradation process called *desertification*.

The bare bones of the Sahel offer a stark lesson in the anatomy of a decades-long drought. Much of Africa has seen periodic ups and downs of rainfall throughout the last forty or fifty years. But the Sahel appears to be caught by some sort of feedback mechanism that prevents its recovery from the short-term droughts that only briefly affect the rest of the continent. Desertification could presumably change the color and tone of the earth's surface and therefore change its reflectivity or albedo. Reduced albedo should lead to reduced soil moisture, convection, and rainfall. Realistic modeling of albedo and soil moisture changes, however, fails to fully account for long-term drought in the Sahel.

A great deal of interest is now focused on dust introduced into Africa's lower atmosphere by wind. The reddish dust rises in stupendous quantities on winds called *harmattans* that have darkened the sky for ages. "Blood rains" from the Sahara fell on Portugal and Spain in 1901; an estimated 2 million tonnes of red mud coated southern Europe in April 1926. Escalating rates of soil disturbance can only inflate these figures. Dust rising to 5,000 m significantly accelerates local atmospheric heat absorption. The high-level African Easterly Jet Stream normally carries rain-bearing clouds to the Sahel from July through September when the ITCZ swings to its northern apogee.

The location of both the ITCZ and the African Easterly Jet Stream depends on regional north-south temperature gradients. When dust is present in sufficient quantities, temperature gradients can shift sufficiently to displace both the jet stream and the ITCZ's moisture-bearing storms away from the Sahel. If this relationship withstands scientific scrutiny, it would be a link connecting at least one type of decade-scale climate change to the actions of humans. Are we comfortable thinking of ourselves as an "external force" that can so fundamentally change the world in which we live?

Oceans 6 and Air

Oceans and the atmosphere inextricably intertwine in the dance we call *climate*. Water and air share common characteristics: Both are complicated mixtures of many different compounds, and both move in patterns determined by landmasses. Motion in both the oceans and the atmosphere is steered by the curveball Coriolis effect generated by the earth's rotation. But air and water differ greatly in how they acquire and transmit heat. Unstirred water conducts heat twenty-seven times faster than air. At sea level, water has a heat capacity 3,500 times greater than an equal volume of air. The oceans, therefore, are vastly more energetic than the atmosphere. The topmost 10 m of an ocean typically holds as much energy as the entire atmosphere above it. The warm oceans between 30° north and 30° south latitude can be thought of as the great thermal flywheels that keep our climate spinning.

Energy is not distributed homogeneously within the oceans, neither vertically nor horizontally. Surface waters at most latitudes are warm relative to deeper waters for a simple reason: Surface water directly absorbs incoming solar radiation; once warmed, this water expands, becomes slightly less dense, and tends to "float" on denser, deeper water. This process is so effective at thermally segregating seawater that most oceans have a clearly demarcated plane, called the *thermocline*, separating their surface and deep waters. At the thermocline, which might lie anywhere from 100 to 1,000 m beneath the surface, water temperatures suddenly plummet by 10 °C or more over a further descent of a few tens of meters. Water above the thermocline can warm and cool seasonally, but deep water will have a temperature, usually 4 °C or less, that is remarkably stable over periods of hundreds of years.

Surface heating also varies horizontally, with temperatures generally increasing as one approaches the equator. Surface temperature becomes disproportionately important as water warms beyond 25 °C. Above this temperature, H_2O is more inclined to exist as a vapor rather than a liquid. A warm band of tropical water encircles the earth, lying within 20–25° latitude of the equator. Because of the earth's tilted axis, the band shifts to one side of the equator or the other, depending on whether summer happens to be in the Northern or Southern Hemisphere. Evaporation from the tropical oceans liberates a vast quantity of water that is carried into the atmosphere. The Tropics are always studded with thunderstorms and regularly breed hurricanes and cyclones. Hot moisture-laden air rising from the Tropics enters the atmosphere and spreads north or south of the equator. Cooler air must slide back in along the surface—converging from north and south—to replace that which has risen within the ITCZ.[1]

Once this oceanic soup begins to boil, it is quickly stirred. Surface currents are set in motion by wind. Global wind patterns are dominated by easterly tropical trade winds, midlatitude westerlies, and polar easterlies, reproduced symmetrically about the equator. A surface current within an ocean will typically achieve a velocity that is about 3 to 4 percent of its generating wind, but will only do so if the wind has blown persistently over an uninterrupted fetch of open water. Antarctica is surrounded by the only ocean where wind and ocean currents girdle the globe without interruption. Consequently, the Antarctic Circumpolar Current transports upward of 200 × 10⁹ kg/s of seawater past the Cape of Good Hope.[2]

Beyond Antarctic waters, however, predominant global winds and ocean currents encounter landmass barriers. Equatorial currents are driven by easterly trade winds that blow unabated across the entire mid-Atlantic and Pacific Oceans: fetches that approach 10,000 km over which the westward-flowing currents can intensify. Inevitably, the currents collide with Asia or the Americas. There they are turned left or right by the continental barriers and forced north or south until westerly winds steer them back across their respective oceans. What is created in each case is a colossal gyre of ocean water, spinning clockwise in the Northern Hemisphere and counterclockwise in the Southern Hemisphere.

Incoming solar energy is most avidly absorbed around the earth's midriff. Here solar radiation strikes the water at angles close to 90° most of the year, minimizing reflectance and maximizing absorption. Ocean gyres redistribute this heat from the equator toward the poles. Smaller counter-rotating currents, such as those in the Labrador and Weddell Seas, transfer warm water even farther toward the poles. The work of redistributing the heat from the equator toward the poles is split about evenly between atmospheric and oceanic currents.[3]

Of course, the picture of ocean currents becomes infinitely more complex when one quits squinting and begins to look closely at any given part of an ocean. Yellow rubber duckies and Nike shoes, lost overboard from storm-tossed ships, have been found by beachcombers on shores around the Pacific Ocean. Curtis Ebbesmeyer and James Ingraham, oceanographers in the Pacific Northwest, have entertained themselves for years by following such objects to track the seemingly whimsical patterns of ocean currents.[4] The resulting maps are best termed "spaghetti diagrams" because of the looping, twisting paths that objects floating in currents follow.[5]

Sea surface temperatures, thermocline depths, upwelling velocities, and upper-ocean currents all vary on a decadal scale.[6] These parameters are usually subtle and are rarely noticed directly by humans; instead we glean insight by reviewing records of ship captains that stretch back a few hundred years. We wade through mountains of data recorded by buoys stationed throughout the world's oceans. We look for anomalies— deviations from the "norm"— to alert us to processes that may be silently operating. As we better understand climate and climate variability, the "norm" is becoming a less reliable benchmark in atmospheric and oceanographic studies. For now, anomalies provide the strongest light in which to observe processes that would otherwise vanish beneath the waves.

Oceanographers have modeled heat anomalies that start near the coast of Japan and extend into surface waters of the central Pacific Ocean.[7] As they move east away from Japan, these pools of warm water spread to both the north and the south. As a result, atmospheric longitudinal temperature gradients over the Pacific are reduced and westerly winds slacken. Driven by less wind, the North Pacific Gyre slows down. The stationary Aleutian Low off the coast of Alaska is weakened, thereby altering North American weather patterns. But the seeds of demise are sown within this climatic field from the outset. The oceanographers' model suggests feedback loops between the ocean and atmosphere, which ultimately account for a twenty-year life span of the initial perturbation.

For those of us of the lumper (not splitter) persuasion, it's not unreasonable to say that ocean currents come in two flavors: surface currents and deep-water currents. The former extend from the surface down to typical depths of a few hundred meters and affect perhaps 10 percent of the earth's ocean volume.[8] Surface currents, propelled by the wind, are molded by an air-sea interaction known as *Ekman transport*. This phenomenon, a spin-off of the Coriolis effect, deflects ocean currents 90° away from the prevailing wind direction.

Tropical warm water in the upper 1,500 m of the Atlantic Ocean migrates northward in the Gulf Stream, transferring heat from the equator to the pole. The heat is absorbed into the atmosphere and whisked eastward, warming northern Europe. Winter cooling results in this tropical water becoming denser as it slips past Greenland; it sinks and flows southward along the bottom of the Atlantic. The deep Atlantic bottom water crosses the equator on its southbound journey, mingling with Antarctic waters south of the Cape of Good Hope.

Deep-water currents are driven by tiny density differences introduced by variations of temperature and salinity. Both surface and deep-water currents can significantly influence climate patterns on worldwide scales, though surface currents appear to be the more volatile of the two and therefore are more likely to be associated with climate variability on a human time scale. In considering deep-water currents, one must keep in mind that the atmosphere cannot respond directly to water that is more than a few hundred meters below the surface. At some point, deep-water currents must force changes at the surface if they are to affect climate.

Oceans can induce climate variability on a longer time scale through

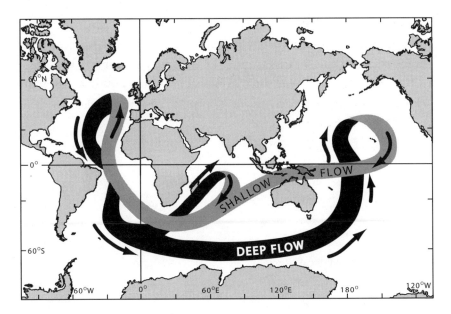

Deep ocean circulation driven by salinity. (Based on Pinet 1998)

a deep-water mechanism known as the *thermohaline circulation*. The fundamental idea is simple: Cold water is denser than warm water, and salty water is heavier than freshwater. The contrasts may be minuscule when measured in a single beaker of water: Differences might span 5°C and 1 gram/kilogram of salt in water. Taken across an entire ocean, however, their effects are staggering (see figure above).[9]

Deep-water processes govern 90 percent of the oceans' volume but are much harder and more expensive to study than surface processes. Wallace Broecker has suggested that this worldwide thermohaline circulation is held in delicate balance with incoming freshwater from the continents that surround the North Atlantic.[10] He describes potential instabilities in the thermohaline circulation caused by incremental changes in freshwater from the Northern Hemisphere. He suggests circumstantial evidence of past reorganizations of deep ocean circulation that could have dramatically disrupted worldwide climate in only a few years.

Oceans are in the business of moving heat. Indeed, half of the earth's heat flux is carried by ocean currents, both deep and shallow. This flux

has enormous ramifications for the world's climate. The oceans and atmosphere are two ends of a dipole that together constitute climate. As the
oceans' share of the heat flux varies—either because winds and surface
currents change or because the thermohaline circulation rolls over in its
deep sleep—the incidence and location of floods and droughts will be
shifted around the globe.

Climate in 7 Hindsight

Climatologists' greatest contribution to society may be their long-term perspective on the earth's atmosphere. Such farsightedness is a state of grace not easily attained. It requires more than just closing the eyes, furrowing the brow, pinching thumbs and index fingers together, humming a proper mantra, and issuing learned pronouncements. Anybody can do that. Instead perspective must be grounded in fact and gained with experience. The facts must be available, must be logically arranged, and must extend far enough back in time. Today the study of climate is facilitated and simultaneously threatened by the sheer volume of data collected daily throughout the world. Under this constant avalanche of numbers, it's all too easy to be caught up in the instantaneous vagaries of weather. What is more difficult—and more worthwhile—is to perceive patterns within this data and to anticipate trends in order to react in a timely manner. To achieve this perspective, one needs a sense of history.

Few climate data series stretch more than a century and a half back from the present, and the ones that do are pretty sketchy. Regular rainfall measurements were first made about 1850; barometric readings and sea surface temperatures were first reliably recorded on a global scale sometime between 1870 and the end of the nineteenth century. These data, maintained at widely separated locations using different methods and mechanisms, were not integrated for meaningful cross-referencing until 1948. This absence of long-term data makes it very difficult to gain a perspective of climate change with historical records alone. To make matters worse, the swings of weather from wet to dry, from hot to cold, have been wide in the past, even wider than what has been observed in just the last 150 years.

Several techniques have been developed to extend the instrumental climate record farther back in time. Each of these techniques has its strengths and weaknesses; each has a time frame to which it is particularly suited and over which its resolution is particularly high. Real perspective on climate emerges when one technique can be used to calibrate and refine another technique. Let's briefly examine a few of them.

Haphazard weather observations have been made and recorded for four or five centuries by ship's captains, lighthouse keepers, and the like. With a bit of imagination, one can find sporadic anecdotal information sprinkled throughout the most unlikely annals of history: grain prices in medieval Europe, calendar dates on which wine grapes were harvested in France, the first mention of spring floods on the Missouri River, or a ship's chance encounter with icebergs at a given latitude in the North Atlantic. Honeysuckle plants, it turns out, have been blossoming a few days earlier each year throughout the western United States. These serendipitous informational nuggets can be cross-checked against other sources, but taken by themselves, they are of limited value. What is needed is a seamless record that can lead step-by-step, year-by-year back into the past.

Historical records yield high-resolution information if they happen to be sequentially intact. Perhaps the most spectacular continuous written record is that of flooding on the Nile River: Reliable information reaches all the way back to A.D. 622.[1] Annual floods were the lifeblood of Egypt. If the river rose to a certain minimum, everyone from farmers to pharaohs knew that the subsequent harvest would be adequate or better as the river inched higher on *nilometers* near Cairo. At first the Egyptians simply scratched marks into cliffs to record a particular year's maximum river height; then they progressed to more sophisticated marble scales, and later to a series of measured steps leading from the river into nearby wells. Joseph himself is said to have built the first nilometer at Bedreshen.

His dream of seven years of floods followed by seven years of famine would have been the world's first long-range climate prediction.[2]

Several natural systems provide much lengthier records of past climates. Tree-ring dating, or *dendrochronology*, offers a valuable proxy record of climate. In temperate and high-latitude locations, trees add annual layers in which width and composition reflect temperature and precipitation at the time of growth.[3] Any one tree might have had a spectacularly good or bad year, but with sufficiently wide sampling throughout an area, the forest of climate variation begins to emerge from the trees. Because the thin-and-thick growth patterns are reproduced faithfully enough among the majority of trees in an area, a reliable one-year resolution of climate condition is possible with the tree-ring method.

By studying cores from trees with overlapping but increasingly greater ages, the dendrochronology series has been extended back to the year 622 B.C. in the American Southwest. By examining the entire tree-ring record, scientists can sense the rhythm to which floods and droughts dance. The rhythm is not a simple syncopation, however. No regular cycles are apparent in the record.[4]

Dendrochronology has yielded a wealth of insight into past climates. Scientists have collected and massaged so much data in the Southwest that they are now able to show the year-by-year and county-by-county advance and retreat of droughts that occurred hundreds of years ago.[5] Fire scars found consistently in trees across Arizona and New Mexico bespeak periods of tinder-dry drought conditions from 1740 to 1760 and from 1840 to 1860.[6] Narrow-growth rings in trees in the central and eastern United States from 1637 through 1641 have been correlated with four volcanic eruptions that opaqued the world's skies and depressed temperatures for one or two years.[7]

Ice, as it accumulates, traps a second independent source of evidence that can also illuminate paleoclimates. Snow falls on ice caps at high latitudes (like Antarctica) and high altitudes (like the Andes of Peru). In these settings, where year-round temperatures remain sufficiently low, ice builds without thawing. The Quelccaya Ice Cap, 5,670 m up in the Andes of southern Peru, has yielded two cores, each a bit longer than

150 m, representing the last 1,500 years of snow accumulation.[8] The cores have been studied in excruciating detail. Volcanic ash from the eruption of Huaynaputina in the central Andes was caught in the A.D. 1600 layer of ice. During the Little Ice Age, dust was deposited at rates 20 to 30 percent above average, suggesting that wind velocities across Peru's Altiplano were stronger than at present.[9]

Scientists have measured oxygen isotopes from eight to twelve points in each annual layer of dust within the Andean cores, teasing out the ratios of "normal" versus "heavy" oxygen (^{16}O vs. ^{18}O) in the ice. Because of subtle differences in weight, the concentration of ^{18}O steadily decreases relative to ^{16}O as the air temperature cools. The ratio between normal and heavy oxygen, referred to as $\delta^{18}O$, thus tracks temperature changes not only annually from winter to summer, but also as the climate shifts back and forth between cold and warm.[10]

Other proxy methods are being developed. Pollen and microfossils dredged from the bottom of Chesapeake Bay record prolonged periods of hypersalinity during the sixteenth and early seventeenth centuries, reflecting the drought conditions mentioned in the diaries of early American colonists.[11] Corals growing within tropical oceans are being measured for $\delta^{18}O$; trace minerals such as manganese, cadmium, and barium can reveal patterns of sea surface temperature and local rainfall that cover periods from hundreds to thousands of years before the present.[12] Annual sediment layers called *varves* accumulating off the Santa Barbara, California, coastline display details of a climate that has changed periodically over the last 45,000 years.[13] Calcite deposited in a groundwater-flooded desert cave has provided a continuous 500,000-year record of climate in the greasewood flats of western Nevada.[14]

What do we learn from paleoclimate records? We see the waxing and waning of climate changes that have occurred in the past, often on scales that dwarf floods and droughts of the last one or two centuries. Sediments perched on the walls of Grand Canyon suggest that a 14,160 m^3/s flood — twice the size of the largest modern events — rushed down the Colorado River about 1,600 years ago.[15] Our 1930s Dust Bowl experience, as devastating as it was, dims in comparison to the tree-ring-recorded drought that desiccated northwestern New Mexico for twenty-one straight years at the end of the sixteenth century.[16] Each of these tools—historical records, dendrochronology, ice cores, coral composition, and cave deposits—are invaluable aids in the search for climatological perspective. As we become more aware of climate change, their value as resources against which to compare the present can only increase.

If the study of paleoclimatology has taught us anything, it's that the earth has gone through episodic warming and cooling in the past. During the early Pliocene Epoch, 4 to 5 million years ago, many locations in the American West were 2–4°C warmer than their modern temperatures.[17] Soon after the last Ice Age retreated 10,000 years ago, sea surface temperatures in the western Pacific Ocean rebounded to levels 1–2°C higher than those of today, before settling down to what has been the average range over the last 5,000 years.[18]

We are still emerging from the chill of the Little Ice Age that ended only 150 years ago. Since then, continental glaciers around the world have been rapidly shrinking. El Niño events transiently increase global atmospheric temperatures, but these are just temporary adjustments as the oceans dissipate their accumulated stores of surplus heat into the atmosphere. If additional heating were to occur, these events could become more common.[19]

The Christ Child

8

I remember driving alone late one January night across northern Nevada, a million miles from nowhere. Entire counties would pass between oncoming pickup trucks. My less-than-trusty Volkswagen—"Otto"—was purring along with the sunroof rolled back and the heater running full blast. There was no moon. The sky would have surely burst if even one more star popped out. Otto's AM radio crackled and hissed as I tweaked the tuner, looking for any voice in that wilderness to keep me awake. I tried frequencies 1400 in Oklahoma, 1070 in Los Angeles, and 770 in Albuquerque—to no avail. Finally San Francisco came in clearly at 680 on the dial. I was good for another hundred miles!

Climatologists have been tweaking knobs too, searching for any and all intelligible frequencies on the climate-change dial. There is tantalizing static at the centuries-long position. Both the Atlantic and Pacific Oceans exhibit rhythmic climatic patterns when measured over decades. But these days the music is clearest when we tune into the three-to-seven-year bandwidth. What we hear is the unmistakable tempo of El Niño.

What is El Niño? That's easy: *El Niño* is a coupled oceanic and atmospheric phenomenon that occurs every few years (roughly every three to seven), with ramifications that ripple out from the equatorial Pacific to touch the lives of people from Peru to Canada to South Africa. El Niño is the quasi-periodic fluctuation of equatorial Pacific water temperatures and the coupled atmospheric response: warm water sloshing back and forth across the basin, with winds rising and falling in unison. You might think of this innate oscillation as the climate system unconsciously drumming its fingers. The devil, of course, is in the details. Perhaps it's best if we begin with a snapshot of the "normal" Pacific Ocean stretching from Peru and Ecuador west to Indonesia.

The general flow of water in the southern Pacific Ocean is dominated by the South Pacific Gyre, a counterclockwise drift driven by global winds and shaped by the Coriolis effect. The eastern limb of the gyre includes the Humbolt and Peru Currents, which flow north past Chile and Peru. These currents are sustained by persistent winds blowing northward along the western coast of South America. As wind pushes the water, the phenomenon known as *Ekman transport*, caused by the earth's spin, dictates that flow will be redirected to the left of the wind track in the Southern Hemisphere. Thus surface water moves away from the coast. Deep cold waters rise to fill the void. The thermocline—that sharp delineation between a warm surface layer and cold deep water—is usually very shallow in these upwelling regions, typically at 40 m or less. Surface-mixing processes reach deep enough to bring up cold water from beneath the thermocline.[1] The take-home message: Pacific Ocean surface water along the coast of South America is normally cold.

Peruvian fishermen, accustomed to these nutrient-rich cold waters that normally upwell along their coastline, had long ago learned to recognize a current of warm water that was likely to change fishing prospects about Christmastime each year. Some years, perhaps one out of every three to seven, the current would be dramatically warmer and the fishing would be much poorer. In 1891 Luis Carranza published a brief note about this current and its associated abundant precipitation in *El Boletín de la Sociedad Geográfica de Lima*. A year later, Camilo Carrillo commented on Peruvian sailors' knowledge of the current they called *the Christ Child*, or *El Niño*, which was strong enough to significantly lengthen or shorten trips up or down the coast in December and January.

Cold upwelling waters along the Peruvian coast, disrupted by the annual incursion of a moderately warm current from the north, is the "normal" picture in the eastern equatorial Pacific. Now let's turn our atten-

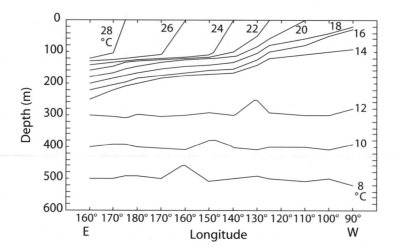

Typical temperature depths of the equatorial Pacific Ocean from Indonesia (160°E) to the western coast of South America (90°W). The thermocline stands out as the band of tightly packed *isotherms* (contour lines of equal temperature). (Based on Pinet 1998)

tion to the western Pacific Ocean near Indonesia. This region is directly influenced by persistent easterly trade winds that converge at the equator from both the north and the south.[2] These winds drive equatorial currents that pile western Pacific water into a mound 60 cm high.[3] The winds tend to diminish as they reach Indonesia; the Southern Equatorial Current slows down there and the water heats up, frequently exceeding 28°C. The thermocline is a couple hundred meters below the surface.[4] The take-home message: The western Pacific is typically a bulging pool of very warm water.

We need one additional element to complete this simplest of models: atmospheric circulation. Waters of the western Pacific approach the limits to which an ocean can absorb solar energy. Above 30°C, evaporation occurs at such a furious pace, and heat is transferred so quickly into the atmosphere, that the ocean cannot long sustain higher temperatures. This air expands and rises, carrying prodigious amounts of both heat and moisture into the troposphere. Much of the rising air moves north and south from the equator in the Hadley cells described earlier. But a significant fraction also flows laterally away from the western Pacific, heading either farther west or back east in a *zonal* (as opposed to a *meridional*, or north-south) flow. Air flowing eastward at upper tropospheric altitudes descends upon the eastern Pacific. Trade winds on the surface complete the loop back to the west.

Cold water dominates the eastern equatorial Pacific, and warm water dominates the west; moist air rises from an area of low pressure in the west, and dry air descends in the east, creating an area of high pressure. This system is held in dynamic equilibrium between solar heating and the earth's attempts to redistribute that heat. The equilibrium drifts back and forth across the equator annually as the ITCZ tracks north or south and the sun moves from winter to summer. You would think that the system would remain within the bounds of this equilibrium, but it doesn't.

El Niño

Gilbert Walker was a mathematician and physicist at Cambridge University before being asked to direct the Indian Meteorologic Service in 1903. Established in 1875, the service had initially been charged with predicting the tropical cyclones that ravaged British ships in the Bay of Bengal. Soon the service was also expected to predict droughts that decimated India when its precious monsoons failed to materialize. More than 4.5 million people had died in the drought and terrible famine of 1899, just before Walker arrived.[5] Ships were sinking and crops were failing as the new century dawned. Directing the meteorologic service was not going to be a breeze.

Gilbert Walker knew little about meteorology when he first showed up, but he was scientist enough to carefully gather data from specific locations throughout India and to simultaneously take a very large view of worldwide weather patterns. Parochial efforts would produce only near-sighted predictions; India needed something better. Walker looked at snow cover in the Himalayas, rainfall over the Nile, and atmospheric pressure records from Australia to Chile. What he most wanted was to create a reliable system for predicting the Indian monsoon. Along the way, he stumbled upon a persistent pattern of oscillating atmospheric pressure:

first low pressure in northern Australia, and then low pressure farther to the east.[6] The oscillation seemed to seesaw back and forth every few years.

The *Southern Oscillation* is now standardized as the pressure difference between Tahiti and Darwin, Australia. When one is high, the other is low. As one begins to fall, the other begins to rise. Walker knew he was onto something, but his correlations of this Southern Oscillation with the Indian monsoon remained less than reliable.[7] It would be left to future scientists to reveal just how large a tiger Walker had grabbed by the tail.

Jacob Bjerknes taught meteorology at the University of California in Los Angeles during the 1960s. He laid the detailed groundwork that was the foundation for understanding zonal air movement over the equatorial Pacific. He called this movement the *Walker Circulation* in honor of Sir Gilbert. In three landmark papers, Bjerknes offered evidence of a coupled ocean–atmosphere flywheel, linking the Walker Circulation with rhythmic changes in sea surface temperatures across the Pacific.[8] He noted that, as easterly trade winds decrease, the western Pacific's warm bulge begins to relax and flow eastward toward South America. With warming of the Pacific Ocean near the international date line, more evaporation occurs there and more energy is pumped into the atmosphere above western South America. The trade winds collapse or even locally reverse, further accelerating and sustaining this breakdown of the Walker Circulation. Monsoons in India and Australia give way to drought. The annual El Niño current that flows southward along the South American coast becomes noticeably stronger and warmer as Ecuador and Peru are hammered by unusually heavy rainstorms and floods.

This coupled ocean–atmosphere process has become known as the *El Niño–Southern Oscillation (ENSO) phenomenon*. Its multiyear rhythms suggest that forces beyond volatile atmospheric dynamics must dictate its waxing and waning. Certainly oceans, with their huge capacity to hold heat, have enough thermal momentum to direct this multiyear periodicity, but why is the process not more predictably cyclic? Is an ENSO event triggered by warm ocean temperatures or by relaxation of the trade winds? Which comes first, the chicken or the egg? The answer, unnervingly, seems to depend on whether one is talking to an oceanographer or a meteorologist. What confluence of circumstances must trigger its onset? Is this an internally driven engine, answering only to the whims of the equatorial Pacific Ocean and its overlying atmosphere, or do external forces exert important influences?

Internal and external forces: How do they interact? The answer to this question remains elusive. One approach involves backing into the problem by asking a related question: Has the Pacific Ocean always experienced ENSO events? If not, when did they start, and what changed in the meantime? W.H. Quinn's historical records clearly show ENSO events as early as the 1500s. Jeffrey Dean's tree rings record ENSOs that occurred as much as 2,600 years ago. Ice-core and coral records show that the ENSO machinery has been steadily chugging along for at least the last 2,000 years.

Scientists have extended their vision back 15,000 years in at least one location by examining debris flows that swept into a lake high in the Andes of Ecuador. Radiocarbon dates were obtained from organic material within individual laminae of mud and gravel. Going back in time, the scientists could document wet/dry cycles with a periodicity of 2 to 8.5 years from the present until about 5,000 years ago. Even farther back, a change becomes evident. Before 7,000 years ago, the debris flows were less frequent, occurring about every 15 years. The lake sediment records end about 15,000 years before the present. The implication is that a clear ENSO-like signal began to develop 7,000 years ago and was not fully recognizable until 5,000 years ago.

What was so different 7,000 years ago? That's a long time back, but even so, some ideas begin to emerge. Holocene molluscan assemblages (older than 5,000 years) along the Peruvian coast north of 10°S are dominated by tropical warm-water species, unlike modern assemblages, which are a mix of warm- and cold-water species.[9] The uninterrupted presence of the warm-water species suggests a continual bath of warmer water along this part of the South American coast. Such conditions would not support the Walker Circulation that is a necessary prelude for ENSO events as we know them. Was global ocean circulation prior to 7,000 years ago dramatically different than now? Other scientists have noted that periodic Milankovitch-like cycles of the earth's orbit during the early Holocene Epoch would have suppressed the warming of solar insolation and thus the ENSO sequence of ocean-atmosphere events.[10] When we consider these determinants of ENSO 7,000 years ago, we see two mechanisms reinforcing one another: changing ocean currents (an internal process) and variations in solar energy (an external process).

At one point, Bjerknes thought that individual ENSO events could be triggered by a breakdown of winds along the Peruvian coast. This would have shut off upwelling, the coastal surface waters would warm, and—voilà!—it would start to rain. Such was not the case, however. After reviewing older records of Peruvian weather, scientists noted that the winds kept blowing from the south, with or without an ENSO event.

Bjerknes's hypothesis has been replaced by a new idea. It now appears that packets of warm water depart the western Pacific bulge and head for South America in what are termed *Kelvin waves*. These waves are herded by the *equatorial wave guide* as they propagate eastward. This "guide" is the phenomenon by which transoceanic waves are forced to stay near the equator because of the Coriolis effect. Each Kelvin wave crosses the Pacific in two to three months, spreading warm water across the surface of the eastern Pacific and further depressing the thermocline. As the eastern thermocline deepens, cold deep waters are pushed beneath the reach of surface-mixing processes. Wind-driven, Ekman-inspired upwelling continues, but it only stirs warm water at the surface; it does not bring up the cold, deep Pacific water, as it does during non-ENSO years. At this point, El Niño is up and running.

Scientists have roughed out an idea of what an average El Niño event looks like, calling this the "canonical" ENSO.[11] In a prelude phase, stronger-than-average easterly trade winds pool warm water in the western Pacific for eighteen months before the inception of an El Niño. Onset is marked in September by a tongue of warm water that extends eastward as the trade winds begin to die. The El Niño event itself begins in December or January as the tongue reaches the South American coast; water temperatures there continue to climb to anomalous peaks between April and June. Sea level along the eastern Pacific equatorial coast rises dramatically. The trade winds reverse, blowing moisture into the otherwise arid deserts of coastal Peru. The high atmospheric pressure, normally so stable over western Peru, deteriorates as the most active portion of the ITCZ moves from the western warm pool eastward into the central Pacific. Rain begins to fall in the Andes, and floods lash down canyons leading from the mountains to the coast.

The eastward-moving bulge of warm Pacific water splits when it hits South America and races both north and south along the eastern margin of the Pacific basin, eventually raising coastal water levels from southern Chile to the Gulf of Alaska. The event enters a mature phase as a second tongue of moderately warm water arrives in the following December, but sea surface temperatures subsequently plummet as conditions return to "normal."

That is what is supposed to happen on average, but no two El Niños are created equal. Some are relatively weak; others are relatively strong. Some start early; others start late.[12] During the past 450 years, El Niños of moderate strength have occurred on average about every four years, and strong or very strong events have occurred about every ten years.[13]

One of the twentieth century's strongest ENSO events occurred in 1982–83. This El Niño took scientists by surprise because it lacked what was thought to be the necessary prelude phase of increased trade winds in the west. The normally westward-flowing South Equatorial Current disappeared in November 1982, to be replaced by the eastward flow of surface waters. Kelvin waves twice raised sea levels in northern Peru and southern Ecuador to 60 cm above normal, and nearshore surface-temperature anomalies rose as high as 10 °C.[14] Peruvian anchovy and sardine stocks collapsed. Seabirds died by the millions. Coastal Ecuador and northern Peru were plagued by flash floods and buried beneath billions of tons of Andean mud. Towns were cut off from the world as roads dissolved and bridges washed away like leaves in a mountain stream. Malaria rose on the wings of mosquitoes, the small but deadly horsemen of this apocalypse.

Huanchaco is a fishing hamlet on the northern coast of Peru. The streets are quiet. Fishermen mend gill nets by their boats on the beach. Some straddle reed boats called *caballitos* amidst the waves, fishing lines held patiently across their bare palms. Others pay a quarter sol to fish from the town pier. The *garua* mist wafts in from the ocean at night, but otherwise there is precious little precipitation. The amount of rain-

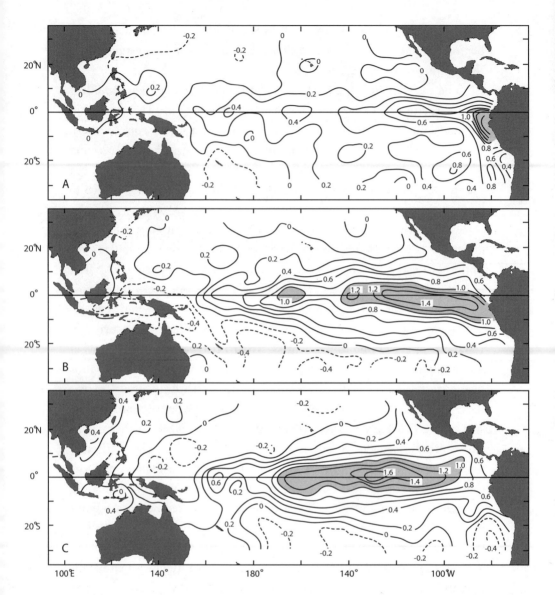

Oceanic response to an average El Niño event: (A) average sea-surface temperature anomalies for March–May; (B) average sea-surface temperature anomalies for August–October; (C) average sea-surface temperature anomalies for December–February. (Based on Cane 1983)

Traditional fishing craft, called *caballitos*, on the beach at Huanchaco, Peru, 1998.

fall from 1943 through 1970 totaled 46 mm.[15] The well-named Río Seco (Dry River) is a dusty arroyo along the south side of the village.

The Moche people occupied this area 1,500 years ago; their ceramics depict caballitos then as they appear now. The Moche mined bird guano offshore and spread it on their fields long before Europeans discovered this agricultural aid. The road from Huanchaco passes by the ruins of Chan Chan, built around A.D. 1300 by the Chimu people who had supplanted the Moche. At 28 km², Chan Chan was and remains the world's largest mud city.[16] Now abandoned, it was once home to 50,000 people. In a land this dry, they were sustained by an extraordinary irrigation system that harvested water from the Río Chicama on the north and from beyond the Río Moche on the south.[17] The main canal system stretched hundreds of kilometers along the coastal plain, always vulnerable to El Niño gashes that would sever its arteries every decade or so. A flood of Noachian proportion, still remembered culturally as Nyamlap's flood, reduced the canal system to rubble around A.D. 1330.[18] Irrigated agriculture here has never again reached the extent attained by the Chimu.

The Christ Child

The Tschudi Complex of the Chan Chan ruins northwest of Trujillo, Peru, 1998.

The Moche and Chimu people no doubt survived many an El Niño. The modern-day people of Huanchaco experienced not one, but two, great El Niños as the twentieth century drew to a close. By mid-1996, the Pacific Ocean beyond 160°W had developed a thick warm layer, deep thermocline, and sea surface temperatures 0.5°C above normal.[19] Conditions, as they say, were ripe. The western Pacific was buffeted by unusually heavy thunderstorms in late 1996 and early 1997. Between trade wind pulses, Kelvin waves were hurled eastward toward South America, toward Huanchaco. By December 1997, another El Niño was underway.

The Río Seco ran bank full in 1998. Unfettered by zoning, homes in Huanchaco that had crept into the channel since 1983 were flicked away like bathtub toys. Air temperatures reached 41°C, a record level in this seaside village. The stifling beaches were littered with rotting dolphins. The Río Seco once again rumbled along the base of the ruins at Chan Chan, rehydrating the Chimu ghosts and further eroding the walls of their mud city.

To the south, the larger Río Moche lapped at the doorstep of Trujillo, a city of 750,000 very damp souls. The Moche is a proper river drain-

A Peruvian family rendered homeless by flooding on the Río Moche upstream from Trujillo, Peru, during the 1997–98 El Niño.

ing 2,700 km²; most of the basin lies above 1,500 m. The river originates high in the Andes, where rainfall exceeds 1,400 mm/yr, sustaining steady year-round flow. In February 1998, the usually tame Río Moche morphed into a dragon. A retaining wall above the city's Mampuesto cemetery failed. As floodwaters sliced through the soft soil, caskets and corpses were unearthed and sent sailing in a grizzly spectacle through the streets of Trujillo.

I talked with a farmer who had abandoned his fields along the Río Moche. Lorenzo was squatting on the outskirts of Trujillo in a cardboard box with his family, just a couple tire widths from the trucks that thunder down the Pan American Highway. He had given up farming, driven out by the floods. How he made a living, I could not tell. With a son in tow, he washed laundry in an adjacent irrigation ditch. Farther upriver, near Laredo, I found Walter living with his daughter and son beneath a piece of fabric in the Río Moche's floodplain, not a stone's throw from the bare foundation of his former home. It had disappeared, swept away by the Río Moche six months before. Walter's naked one-year-old son stared at

The Christ Child

75

Lorenzo and his son, squatters along the Pan American Highway outside Trujillo, Peru, 1998.

me, his gaze as relentless as the sun. Walter's features were hawklike, defiant. He would rebuild his home on its old foundation. After all, how else could he sell chicle to passing trucks and buses on the road that would be rebuilt in this floodplain? How else could he feed his family? Men nearby, armed with only shovels and smiles, were working on a washed-out bridge a little ways downstream. What else could they do?

Fishermen in Salaverry, a few kilometers to the south, had seen El Niños come and go before. After all, their forefathers had named these currents a century or more ago. Antonio Huamanchumo unloaded manta rays and a fish called *tollo* on the Salaverry pier. Earlier in the year he had less to show for his efforts: The warm waters of 1997–98 had decimated his fishing grounds. He simply quit fishing. He had spent his days attending to his beached boat as it dried and cracked, waiting for the waters to cool and the fish to return. Antonio was uncertain about a connection between the warm waters and the unusual rains. He was a fisherman, not a farmer, and said that he hadn't thought about those things.

When Kelvin waves collide with South America, they turn the cor-

A fishing boat beached during the 1997–98 El Niño at Salaverry, Peru.

ner and head both north and south away from the equator. Farther from the equator, where the Coriolis effect is stronger, some of this energy bounces back across the Pacific as *Rossby waves*. At higher latitudes the migrating bulges of warm water become trapped by the increasingly compelling Coriolis effect, turning them to the right in the Northern Hemisphere and to the left in the Southern Hemisphere. The result is that beyond the tropics of Cancer and Capricorn, the poleward-racing warm water is pushed against the coast of Chile to the south and against central America, Mexico, and the West Coast of the United States to the north. The atmospheric consequences of this redistribution of warm water spread pent-up heat from the equatorial Pacific throughout the world. El Niño events are one of the workhorses by which the earth redistributes excess solar energy from the equator toward the colder poles.[20]

At the peak of the 1997–98 El Niño, sea surface temperatures in the eastern equatorial Pacific reached 5°C above normal. By May 1998 the event collapsed, and temperatures plummeted an astounding 1°C every week. By June surface temperatures at 110°W longitude were 3°C below normal.[21] La Niña, the cold event, had arrived.

La Niña

La Niña, at first glance, is simply the inverse of El Niño: Trade winds are enhanced, the western Pacific is particularly warm, and cold upwelling waters return to the coast of Peru even more than usual. In many places around the world, La Niña reverses the atmospheric consequences of El Niño events: dry weather where it had been wet, wet weather where it had been dry. The La Niña that occurred during the spring and summer of 1988 was a prime contributor to the North American drought that carried a $40 billion price tag. There are divergences from the inverse relationship between El Niño and La Niña because La Niña chilling of seawater can suppress evapotranspiration only so far (that is, toward zero). El Niño heating, however, will have an ever-increasing effect as surface waters approach or surpass 30°C.[22]

In the southwestern United States, there is one very important functional difference between the effects of El Niño and La Niña. Although floods are the expected consequence of El Niño in Peru, the average El Niño could cause either floods or drought in the Southwest. This may have something to do with that difficult-to-define strength of the El Niño: strong versus moderate or weak. Even the 1997–98 El Niño, which some called the strongest of the twentieth century, did little in the Southwest besides pushing rainfall above average. Drought, however, is the reliable consequence of La Niña in this region. While not all El Niños are created equal, the effects of most La Niñas are, and the reasons are unexplained.

We see an emerging picture of the Pacific Ocean, with sea surface temperatures fluctuating at the equator from warm to cold to warm again. ENSO events—both warm (El Niño) and cold (La Niña)—are intrinsically coupled oceanographic and atmospheric phenomena. When waters throughout the equatorial Pacific are heated or cooled, the atmospheric effects ripple well beyond that pond. Everything that goes up must come down: Eventually all El Niño events sputter and fail. The simplest explanation for cessation is that the warm Pacific pool exhausts its supply of excess heat. The particular mechanisms of collapse, however, are no better understood than the specific triggering mechanisms that light an ENSO fuse in the first place.

Teleconnections 9

Drought in Brazil and South Africa; cold winters in Tennessee; floods in Arizona and California; tornadoes in Florida; high temperatures in the Sahel and New Delhi. All of these tend to occur simultaneously during strong El Niño events when the Southern Oscillation Index (SOI) is negative.[1] The converse is also true: La Niña tends to create drought in the Southwest, floods in the Northwest, and floods in Australia.

Sir Gilbert Walker was the first to suspect that climatic events were connected on a worldwide basis. Jerome Namias more formally defined the term *teleconnection* in the 1960s to describe a primary climatological disturbance in one corner of the world coupled to secondary disturbances many thousands of kilometers away. Warming of the central and eastern equatorial Pacific, as we have seen, is directly responsible for climate changes from Australia to Brazil.

Teleconnections have not always been obvious. Without a conceptual framework to underscore their existence, ENSO-related teleconnections remained elusive for centuries. Meteorology would struggle well into the twentieth century to get past the point of just licking a finger and sticking it up in the breeze. Biology had entered its modern era with the evolutionary inspirations of Charles Darwin in the nineteenth century. Geology saw itself reinvented in the 1970s in the light of plate tectonics. Today the idea that global links exist between widely separated locations is revolutionizing the atmospheric sciences. Climatologists are rapidly reevaluating their world in view of this new understanding of teleconnections.[2]

North America

The teleconnection web that links global climate is spun from both oceanographic and atmospheric inputs and responses. The eastward shift of the Pacific ITCZ affects not only the east-west Walker Circulation, but also meridional flow of moisture and heat out of the western Pacific. A rise in equatorial Pacific sea-surface temperatures will deepen the semipermanent Aleutian Low, sparking a rearrangement of the jet stream that typically wets portions of North America during winter.[3] During an El Niño event, a subtropical jet stream breaks away toward the south and becomes a conveyor belt bearing the succession of storms that batter the California coast and bathe the Southwest while Washington and British Columbia remain relatively dry.

Dan Cayan is a USGS climatologist who works at Scripps Institution of Oceanography in La Jolla, California. His work has largely focused on streamflow in the western United States as a function of climate perturbations. Cayan remarks that El Niño opens the door to winter storms in California and the rest of the Southwest. *Opens* the door. Storms may or may not come in during a particular El Niño, but they are much more likely then than during a La Niña. For instance, nine of eleven El Niño years in southern California had above normal rainfall, and eight of the ten wettest years from 1950 to 1982 were El Niño years.[4] Cayan's notion of statistical probability is crucial to an appreciation of ENSO teleconnections. Floods in the Southwest can happen at any time, regardless of the ENSO phase. They are, however, much more likely to occur during El Niño years and very unlikely to occur in La Niña years.

The American Southwest certainly feels these ENSO effects. The California shoreline was subject to increased sea cliff erosion and storm damage during both the 1982–83 and 1997–98 El Niños.[5] During an ENSO warm phase, the Southwest is likely to see not only more winter storms, but also greater precipitation totals, more days during which pre-

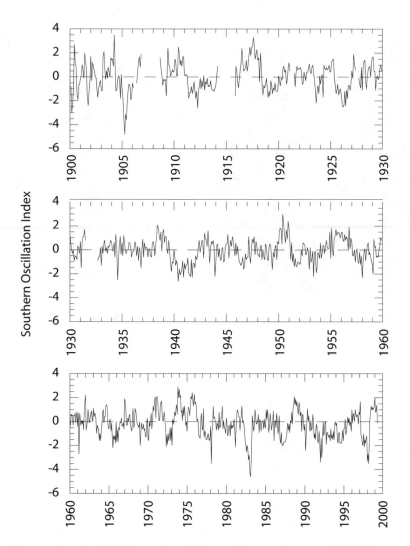

The standardized Southern Oscillation Index (the difference in sea-level pressure between Darwin, Australia, and Tahiti), 1900–2000. Sustained periods of negative SOI are El Niños.

Homes built on coastal bluffs near Bodega Head in California were threatened by intensive wave action by storms in the winter of 1998.

cipitation occurs, and greater streamflow.[6] All of these factors combine to make flooding more likely.

This likelihood of flooding stretches east across the deserts of Arizona and New Mexico. Flow within the Santa Cruz River near Tucson, Arizona, has exceeded its *hundred-year flood (HYF)* level three times in the last four decades, justifiably encouraging the lay public to ask just what the gosh darn heck is going on down there.[7] Part of the problem is that the HYF is a definition, not a prediction; it is the magnitude of flow that has a 1 percent chance of being equaled or exceeded in any given year. The timeworn HYF concept assumes that (1) any one year has an equal probability of flooding as any other year; (2) the time period on which the flood frequency estimate is based is representative of all other time periods; and (3) factors within the basin that affect flooding (such as vegetation densities, urban drainage systems, and river channelization) have not changed. These assumptions might have marginal validity in an area that experiences only small variations in annual rainfall and that isn't busting

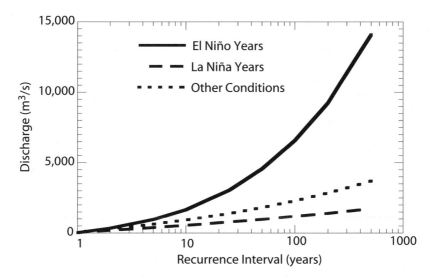

The effect of El Niño and La Niña on flood frequency of the Gila River near Safford, Arizona.

its buttons with new development. But none turn out to be true along the Santa Cruz River.

The first HYF estimate for the Santa Cruz River was 355 m³/s, a calculation made in 1937 based on what was then a twenty-year record of flows. The middle of the twentieth century saw few strong El Niño events and no significant floods, thus reducing early HYF estimates. Subsequent recalculations sequentially increased the HYF level each time another big flood occurred, finally reaching 1,050 m³/s when the 1983 flood was included. Tucson has grown exponentially over the last half century, and the surrounding desert has largely been bulldozed and paved over. The Santa Cruz River is a different river now than it was a hundred years ago, and floods have a much higher potential for wreaking havoc on floodplain structures.

The notion of an unchanging HYF is gradually yielding to a more dynamic understanding that extremes of climate can and will drift over periods of years, decades, and centuries. The biggest change in our thinking about the behavior of rivers is the incorporation of an awareness of variable climate processes. Recurrence intervals of peak discharges on the Santa Cruz and nearby Gila Rivers have been plotted, clearly showing that floods are likely to be two or three times greater in El Niño years than in other years.

Autumn floods in California, Arizona, and New Mexico often occur in the aftermath of hurricanes spawned in the ENSO-warmed waters off the western coast of Mexico.[8] Unusually large dissipating tropical cyclones were parked over the Southwest during the ENSO years of 1925–26, 1939, 1957–58, 1976–77, and 1982–83. When upper atmosphere winds shift between north-south *(meridional)* and east-west *(zonal)* patterns, loops of low pressure can be cut off and left to drift beyond the tug of the jet stream. These cutoff lows, particularly when they suck in tropical moisture left by hurricanes, have long been known to be responsible for some of the paroxysmal precipitation events that occasionally pelt the Southwest.[9]

To the east, Florida also experiences ENSO-influenced weather patterns. Unlike the American Southwest, moisture in the Southeast is drawn from the Caribbean Sea and Gulf of Mexico rather than the Pacific Ocean. Storms are like tanker trucks bearing a load of water over well-worn paths. They might pick up a load of moisture from the Pacific if they're headed toward the Southwest or a load from the Gulf of Mexico if they're bound for Florida. El Niño pushes both storm tracks farther south than usual, simultaneously affecting these two regions even through they border separate oceans. El Niño teleconnections are, in the largest measure, atmospheric rather than oceanographic.

During an El Niño event, the tropospheric jet stream is intensified and develops a southern branch that flows across the southern tier of the United States.[10] Sometimes the path of the subtropical jet stream becomes frozen, and storm after storm travels this freeway over the continent. "Unusual" weather then becomes the norm as this southerly jet stream introduces new air masses to the region. Tornadoes are spawned where dry air blankets an underlying layer of unstable moist air.[11] A tornado with a mouth 1 km wide and winds in excess of 400 km/hr mauled Kissimmee, Florida, during the worst of the 1997–98 El Niño. Rivers like the Suwannee of Florida and others throughout the Southeast tend to

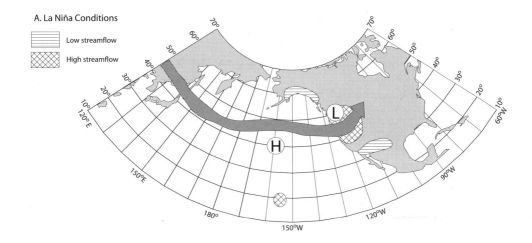

A. La Niña Conditions

▤ Low streamflow
▨ High streamflow

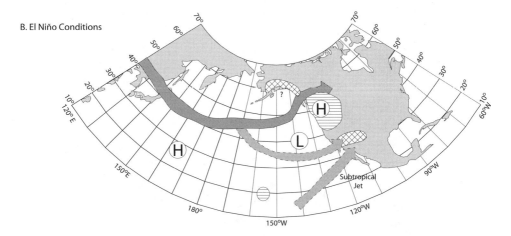

B. El Niño Conditions

Typical winter storm tracks during (A) La Niña and (B) El Niño conditions. (Based on Cayan and Webb, 1992)

rise during El Niño years. This El Niño relationship, though, is less robust than the one observed on rivers in the Southwest because of other confounding factors, such as higher base-level precipitation and tropical storms that sweep up from the Caribbean.

If flooding is the yin of El Niño, drought is the yang. When the standing Aleutian Low is reinforced during a warm ENSO event, the jet stream over North America splits: a weak branch flows north into Canada's

A tornado near Union City, Oklahoma, on May 24, 1973. (Photograph courtesy of NOAA; www.photolib.noaa.gov)

Tornado damage in Kissimmee, Florida, 1998.

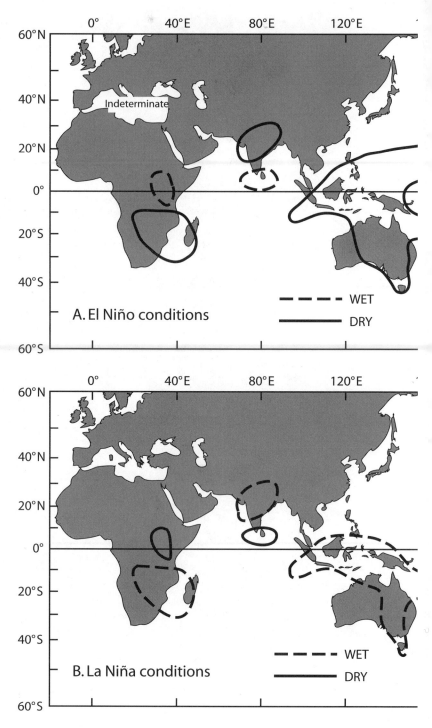

Teleconnections in North America, showing areas of significant positive and negative correlation of seasonal precipitation during El Niño and La Niña events:

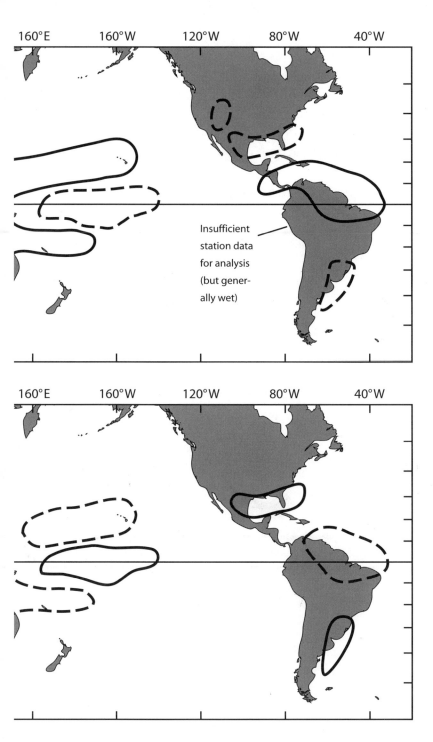

160°E   160°W   120°W   80°W   40°W

Insufficient
station data
for analysis
(but gener-
ally wet)

160°E   160°W   120°W   80°W   40°W

(A) El Niño conditions; (B) La Niña conditions. (Based on Ropelewski and Halpert 1989)

Northwest Territories, and a strong branch flows south, bearing down first on California, then on Florida. British Columbia and the prairie provinces of Alberta and Saskatchewan are left high and dry, out of the moisture loop.[12] Ask the farmers around Hanna about these times of El Niño–inspired drought. The Pacific Northwest and the Missouri River basin experienced extreme droughts during the prolonged El Niño period from 1990 through 1995.[13] There is good, if disturbing, evidence that droughts became more intense worldwide during the last quarter of the twentieth century, as ENSO factors collided with an overall warming of the globe.[14]

Australasia

In 1983 Australian agricultural production dropped by 18 percent, and farmers had 45 percent less income at the end of the year. Their fortunes paralleled the decline of the Southern Oscillation Index that heralded the oncoming El Niño. Australia's crops withered.[15] To be sure, Aussies are an independent lot: Not all of their droughts can be neatly corralled into correspondence with ENSO events. Drought conditions do tend to be much more widespread during those years, with northern and southeastern Australia being the hardest hit.[16] These regional droughts in Australia extend north to Indonesia and Asia.[17]

From Down Under, we go Up Top, to the world's highest peak. It's early May on Mount Everest. The great mountain juts into a jet stream, which all winter has been howling west to east. Climbers huddle at their base camp, sipping tea and shivering, waiting for a break in the weather. Winds at the summit gust at well over 100 km/hr. The climbers watch lines of snow streaming off the peaks like Tibetan prayer flags. It's as if the earth has been inhaling for months. There will be a slim moment suspended in time—a calm day or two when survival is possible at the top of the world—before the winds reverse, and the earth begins to exhale. That is the instant the climbers have been waiting to seize.[18]

The Asian landmass, centered on the Himalaya Mountains, heats as spring and summer warming occurs. Being continental, the land is able to absorb heat more quickly than the nearby Indian and western Pacific

Oceans. During winter, heavy cool air descends from the mountains. Inhale. With the coming of summer, the air heats and the atmospheric tide turns. Exhale. Climbers clatter out of their tents and rush the top of Everest before the easterly winds herald another season.

As the air over India and the Himalayas heats during spring, it rises, drawing moisture from the surrounding oceans. Moist air rising over hot rock: This is the perfect setup for convective thunderstorms of epic proportion. This rising air is the source of the Asian monsoon. No other place on the earth boasts a greater rainy season. Cherrapunji, in the Assam Hills of India, has received 25 m of rain (yes, 82 *feet*) in a single year, arriving mostly during the four-month monsoon. The monsoon occurs seasonally rather than year-round because it is tied to the northward summer migration of the ITCZ.

The East Asian monsoon is conspicuously influenced by two factors: one oceanographic, the other terrestrial. When the western Pacific is warm and the overlying atmospheric pressure is low, increased evaporation supercharges the monsoon with extra moisture and latent heat. When the western Pacific is cold, the monsoon is suppressed. In other words, the Southern Oscillation and the associated phase of El Niño appear to influence the monsoon, thus directly impacting the millions of Asian farmers who rely on its bounty. This is the sort of connection that Sir Gilbert Walker suspected but could not conclusively prove in the early 1900s.

The terrestrial factor may be the amount of snow cover in the Himalayas and on the Tibetan Plateau. Snow, with its high albedo, can prevent continental warming, thus delaying or diminishing the strength of the Asian monsoon. This idea was first promulgated by one of Walker's predecessors, Henry Blanford, at the Indian Meteorologic Service. Blanford had noticed a north-south atmospheric pressure oscillation between India and Russia and had tried to use these data to forecast failure of the Indian rainy season. He suspected that year-to-year variations in snow cover helped to drive the oscillation. The idea has been in and out of vogue ever since. Its most recent incarnations attempt to correlate Himalayan snow-cover surplus or deficit not only with monsoon strength, but also with the onset and maturity of entire ENSO events.[19] The jury is still out.

A spectacular bloom of winter wildflowers in the Desert Lily Preserve in southern Arizona. The bloom followed above-average rainfall associated with the 1997–98 El Niño.

Give and Take

El Niños give, but El Niños also take away. In 1982–83 and again in 1997–98, California and the American Southwest were inundated with precipitation. At the same time, the Pacific Northwest scorched under drought conditions. In general, if rain falls more heavily in one place, it will likely fall less heavily somewhere else. If El Niño is an open door to storms in southern Arizona, it is a closed door in Alberta's Palliser Triangle.

Interestingly, total precipitation in the American Southwest and Northwest does not change much from year to year. If one area is wet, the other tends to be dry. The year 1983 was exceptionally wet, with an overall increase of 30 percent above normal total rainfall, while 1977 was dry, with 20 percent less than normal total rainfall. The annual sum of precipitation in the Southwest and Northwest usually varies by no more than 10 percent in most years, regardless of El Niño, La Niña, or in-between conditions.[20] One person's flood is often another person's drought.

Drought conditions during an El Niño have been documented particularly well in northeastern Brazil. At the height of an El Niño, atmospheric pressures are lower than usual over the eastern Pacific, while pressures over the western Atlantic and northern Brazil tend to be higher than usual as air subsides onto the region.[21] Evaporation from the Amazon basin slows. As a result, Forteleza and the Nordeste of Brazil can experience debilitating droughts during ENSO events.[22] Northeast trade winds in the equatorial Atlantic Ocean are suppressed.[23] The air wallows without wind; evaporation slows as saturation limits are reached. The Atlantic Ocean, deprived of this effective cooling process, steadily warms as the summer sun shines. Because of this teleconnection and others, the effect of ENSO spreads from the Pacific to the Atlantic and thus propagates to a truly global phenomenon.

Hurricanes 10

Atlantic hurricanes, usually spawned off the coast of Africa, rumble west and mature into monsters in the Caribbean Sea and Gulf of Mexico. During El Niño events, hurricane development is suppressed in the Atlantic Ocean because displaced jet stream winds tear the tops off of developing cyclones. Energy that would have been dispersed into the atmosphere builds up in the ocean instead. But the lid popped off as La Niña conditions became established during the summer of 1998. Ten named storms churned through the Atlantic during the thirty-six days from August 19 to September 23. Four hurricanes could be seen in a single satellite image taken on September 26, 1998.

Post–El Niño Caribbean heat dissipation found its most ferocious expression in a storm that ripped through Central America in the fall of 1998. Hurricane Mitch was first detected as a tropical depression with 30 knot winds in the Caribbean (11.6°N, 76.1°W) on October 22, 1998. For a week it meandered north and then west, drawing ever more strength from the warm tropical water before suddenly veering south to slam into Honduras. Atmospheric pressures at the storm's eye dropped to 905 millibars, the fourth lowest of any Atlantic storm in the twentieth century. Mitch had mushroomed into a Category 5 hurricane, with winds of 287 km/hr sustained for thirty-three hours. Gusts exceeded 370 km/hr.

The storm crawled cruelly over Honduras and northern Nicaragua,

A satellite view of Hurricane Mitch northeast of the coast of Honduras on October 26, 1998. (Photograph courtesy of NOAA; www.osei.noaa.gov)

where the mountains' orographic effect wrung a maximum of moisture from the all-too-willing clouds. Rain fell without relief: more than 600 mm in a single six-hour period in Honduras, with six-day totals approaching 900 mm. The Río Aguan in northern Honduras peaked just below 20,000 m³/s. A devastating mudflow swept down the southern flank of Casita Volcano in northwestern Nicaragua. The towns of Rolando Rodríguez and El Porvenir vanished; at least sixteen hundred people died.[1] On the Honduras-Nicaragua border, five hundred bodies were entangled along the banks of the Río Coco; at least another five hundred were washed into the Pacific.

Tegucigalpa, the capital of Honduras, lies within an enclosed basin on the upper Río Choluteca. The city's 1 million inhabitants live in homes that cling to steep slopes above the river. In the last days of October, the river raged. No one is certain how much water flushed through the city because all of the river's gaging stations were washed away. Months after the storm, hydrologists back-calculated the peak flow of the Río Choluteca at Tegucigalpa to be 4,360 m³/s; this volume tripled by the time the Choluteca passed the downstream town of Apacilagua. An entire neighborhood, Colonia Soto, melted like granulated sugar as the Choluteca gnawed beneath its precarious footing, triggering the 6-million-m³ El Berinche landslide.

As early as November 3, the destruction of thirty-three bridges had already been confirmed; another seventy-five were damaged. Rare was the road that was not dissected by tributary floods and debris flows. Seventy percent of the crops were destroyed in this country that lives hand to mouth. The storm's $5 billion in damages set the region back by decades. The disaster was so staggering in its enormity that exact numbers will never be determined. The best guess for Honduras alone is 5,657 people killed, 8,052 missing, 11,762 injured, and 1.9 million impacted.[2] In all of Central America, the storm toll was 9,021 dead and 9,195 missing. This makes Mitch the third or fourth deadliest hurricane known, though its toll pales in comparison to the 1780 hurricane that killed more than 20,000 people in the eastern Caribbean.[3]

Before Mitch, poverty was merely an abysmal fact of life in Tegucigalpa, a curious pungency never far from one's nostrils. Twenty percent of the country's 6 million residents were left homeless by the storm. After Hurricane Mitch, poverty became asphyxiating, wafting through the air like a dark gaseous cloud. Malaria, dengue fever, leptospirosis, and cholera made their morbid rounds. The city has historically had one sewer system: the Río Choluteca. The municipal water system, suspect to begin with, was nonexistent immediately after the storm. Rescue workers in Mitch's wake refused to step in the Río Choluteca for fear of contracting hepatitis.

The name *hurricane* comes from Huracán, the name of the Carib god of evil. Hurricanes begin their tumultuous lives at or near the north and south edges of the ITCZ in all of the world's oceans that touch the Tropics. All hurricanes travel in a westerly direction, steered gently by the easterly trade winds. They begin life as a *tropical depression*, defined as a low-pressure system with maximum sustained winds of 62 km/hr. They escalate to a *tropical storm*, with sustained winds of 62–118 km/hr.

The El Berinche landslide in a neighborhood of Tegucigalpa, Honduras. This landslide was activated during intense rainfall associated with Hurricane Mitch in October 1998.

Children in Tegucigalpa playing among homes destroyed by Hurricane Mitch that were near the Río Choluteca, 1998.

The eye wall of a hurricane seen from a NOAA hurricane-tracking aircraft. This eye wall consists of a vertical bank of clouds thousands of feet high. (Photograph courtesy of NOAA; www.photolib.noaa.gov)

When the storm becomes a *hurricane*, with sustained winds of greater than 118 km/hr, another five categories kick in, culminating in Category 5 storms with winds greater than 248 km/hr.

On a climatological average, 9.8 tropical storms, including 5.8 hurricanes, course through the Atlantic annually. Even more—16.4 tropical storms and 9.2 hurricanes—form off the western coast of Mexico, but most of these travel harmlessly out to sea. By the end of the 2000 hurricane season, only four Category 5 hurricanes were known, and two—Hurricanes Georges and Mitch—occurred in 1998.[4] The year 1999 brought another record, a total of four Category 4 hurricanes, or roughly 20 percent of the Category 4 hurricanes known to have occurred in the twentieth century. In our changing climate, are hurricanes becoming more intense, or can we just measure them better?

Hurricanes in the Atlantic, and typhoons in the western Pacific: Each carves its name into a coastal swath with a fury not otherwise known in the annals of meteorology. They include the deadly Galveston storm of 1900, the hurricane that swamped New England in 1938, and Mitch, which ravaged Honduras in 1998. In the Atlantic, 96 percent of major hurricane days are crowded into a season that stretches from August through October, with a peak in mid-September.[5] This is the time when all of the ingredients necessary for a great storm are present. The ocean has had all summer to heat. Numerous tropical thunderstorms are available to form the nucleus of a hurricane. High-level tropospheric winds have calmed and rarely shear the tops of developing hurricanes. And spinning forces from monsoonal troughs in the Atlantic atmosphere are strong enough to fuel the movement of a nascent hurricane.

William Gray directs a crack team of hurricane forecasters at Colorado State University.[6] In the past decade and a half, Gray has had exceptional success in predicting the number of tropical storms that will occur within the Atlantic during a given year. He breaks his estimates into three categories: named storms, hurricanes, and intense hurricanes. Gray looks at the ingredients listed above and salts them with additional last-minute information about the status of El Niño/La Niña, African Sahel rainfall, and the *Quasi-Biennial Oscillation* (QBO; an equatorial wind that blows at altitudes above 16 km, first from the east, then from the west, in 26–30-month cycles). His success in predicting hurricanes suggests that Gray may have put his finger on the pulse of storm genesis. He's still working out the details, such as where an individual storm will make landfall. As the Hondurans know, this makes all the difference in the world.

The Slow Wheels of Change

11

With an understanding of ENSO and its teleconnections, we begin to see surprisingly far into the realm of atmospheric processes. The lens that focuses this vision relies on a massive network of oceanic monitoring equipment and thousands of weather stations throughout the terrestrial world. We polish the lens with the hindsight of history and aim it with incredibly complicated mathematical models that attempt to simulate real world climate. What do we see? The rhythmic year-by-year breathing of large portions of the atmosphere: now hot, now cold; now wet, now dry; floods one year, drought the next.

Details escape this focused vision, however, on both ends of the spectrum. No one can say that a single storm was generated by ENSO processes; they only provide the statistical framework within which we sagely nod or frown as events unfold. Viewed from the perspective of decades rather than days, some periods show precipitation patterns dominated by El Niño or La Niña events, while other periods exhibit little or no effect by ENSO on the weather. Climatologists are now fascinated by newly emerging observations of decadal-scale variation superimposed on the interannual ENSO signal.

The 1939 sockeye salmon catch in Bristol Bay, Alaska, was the greatest up to that point in history. The 1972 season at Bristol Bay was a disaster, but the fish returned in record numbers in 1995. The pattern for chinook salmon was exactly reversed along the Columbia River in Washington and Oregon: poor fishing in 1939, record gillnetting in 1972, and a dismal catch in 1995.[1] Of course many factors could influence fish populations, but the regularly reversing pattern is just too tantalizing. Could the food chain—beginning with phytoplankton and ascending through salmon all the way to humans—be responding to subtle but real determinants of ocean productivity?

Tree rings along the Baja and southern California coast tell a story that stretches back to A.D. 1660. The plot is simple: About every twenty-two and a half years, Jeffrey pine and big-cone Douglas fir switch back and forth between high-growth and low-growth phases.[2] The two most recent phase shifts occurred in 1947 and 1977.[3] Across the Pacific, tree ring records in Japan suggest that temperature regimes changed there also in 1925, 1947, and 1977. Seasonal streamflow in selected regions on both sides of the equator show not only the expected reaction to individual ENSO events, but also a longer-term pattern of diminished teleconnections from 1920 to 1950 and accentuated teleconnected responses from 1950 through 1980.[4]

The North Atlantic has its own oscillation, with atmospheric pressures inversely rising and falling in Iceland and the Azores. Two and a half centuries ago, a missionary named Gaabye observed that temperatures in northern Europe and Greenland varied inversely. This relationship has subsequently been formalized as the *North Atlantic Oscillation* (as opposed to Walker's Southern Oscillation).[5] Short-term components of this oscillation cycle about once every decade, twice as frequently as does the North Pacific's twenty-year cycle. Longer-term components are observed at forty- to fifty-year intervals, possibly reflecting some form of input from the Atlantic's thermohaline circulation.[6] So far, little conclusive evidence has surfaced that would directly link the oscillations of the Atlantic and Pacific Oceans.

The temperature of the central Pacific Ocean north of latitude 20° has varied about 0.5°C about every two decades. The last sudden change from warm to cold was in 1977. Before that, the temperature had changed from cold to warm in 1947. Half a degree may not seem like much. However, if that same absolute quantity of heat were not spread throughout the North Pacific Ocean but concentrated just in the rivers of North America,

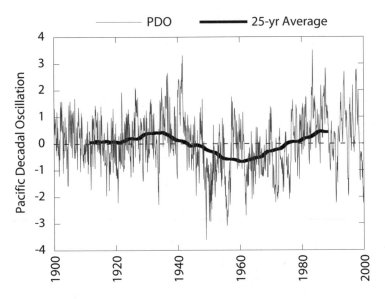

Pacific Decadal Oscillation, 1900–2000. The solid line shows the twenty-five-year running average. (Based on Mantua et al. 1997)

they would all boil away in a heartbeat. Trust me: Half a degree in the North Pacific is a lot of heat.

Something subtle is going on here. Scientists rely on their statistical microscopes to see much of the detail of decadal variation. The trend may be obscured by other energetic processes, but it's real.[7] Climatologists have named this fluctuation the *Pacific Decadal Oscillation (PDO)*. It's said to exhibit a positive or "high" state when the central North Pacific is relatively cold, and a negative or "low" state when that ocean is warm. When the central Pacific is cold (i.e., positive PDO), the eastern boundary—along the western U.S. and Canadian coasts—is anomalously warm.

Associated with a positive PDO is a deepening of the Aleutian Low, nestled in the crook of Alaska's Aleutian Peninsula, where so much of North America's winter weather is generated. With an intensified Aleutian Low, storm tracks are deflected farther south, away from the Pacific Northwest and into the southern tier of the United States.[8] The PDO was mostly positive from 1925 until 1947, mostly negative from 1947 until 1977, and mostly positive after 1977. There are indications that the PDO

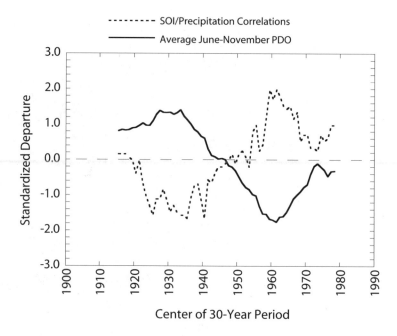

The interaction of PDO and SOI indices. (Based on McCabe and Dettinger 1999)

may have changed again in the waning years of the twentieth century, possibly as early as 1996.

The PDO appears to be driven by forces residing within the North Pacific, while ENSO is a tropical phenomenon with extratropical ramifications. The decades-long memory of the PDO surely resides within the incredible thermal inertia of the Pacific Ocean, rather than just within the more volatile overlying atmosphere. Scientists are still a long way from a single convincing explanation for the specific ocean/atmosphere mechanisms governing the PDO. Possible factors include midlatitude ocean gyre and current strengths, wind-driven upwelling, oceanic or atmospheric input from the Tropics, or some combination of these factors.[9] Whatever the force, it must be subtle and infrequent.

Do the PDO and ENSO interact? Certainly so, if we are talking about their effects on precipitation, flooding, and drought.[10] When an El Niño arrives off the coast of California during a high PDO phase, you can put good money on the probability of flooding in the Southwest or drought in the Northwest. Conversely, when La Niña conditions exist during a low PDO phase, you'd better invest in umbrellas in Seattle and suntan lotion in Tucson. During these times, the two processes are said to exhibit *constructive interference*, reinforcing the flood or drought tendencies of El Niño or La Niña in a given region.

The PDO and ENSO combine to deliver more precipitation to the Southwest when the PDO displaces the Aleutian Low storm track toward the south, and when ENSO delivers increased moisture to these storms as they reach California. If the tables are turned, however, with El Niño during a low PDO or La Niña during a high PDO, all bets are off.[11] During these times of *destructive interference*, it is difficult to predict what weather patterns the ENSO teleconnections will produce.

Decade-scale climate processes don't often make the ten o'clock news, but they do make a difference. The Colorado River Compact of 1922 apportioned that river's flow among seven very thirsty states, based unwittingly on three decades that happened to have been wetter than any that followed during the next half century. The seven states have not quit squabbling since then. Arizona bristled badly enough in 1934 to launch a rickety navy to defend its watery interests when California wanted to dam the lower Colorado River just north of Parker.[12]

Decadal changes can also offer pleasant surprises. Groundwater levels in some parts of eastern Nevada have been rising over the last forty years, despite increased withdrawals in the vicinity.[13] It's difficult to anticipate all the twists and turns of climate, but dams and water-delivery systems are ideally built with the demands of long-term water supply in mind. Roger Pulwarty, a NOAA climatologist, was wise to note that we can *manage* our hydrologic assets for interannual variation, but we must *plan* for decadal variations.

Living in a Web of Climate

Climatic fluctuation can shape the lives of humans, plants, and animals alike. In 1983 corals of Tarawa Atoll (1°N, 173°E) were bleached by stressful salinity changes caused by intense rainfall that occurred when the ITCZ was displaced eastward from Indonesia.[1] Delicate ecological balances were quickly tipped during that major ENSO event. Crabs and shrimp that normally live in obligatory symbiosis with pocilloporid corals in the Galapagos Islands declined as the corals were stressed. The crabs and shrimp, in turn, could no longer protect the corals from invading sea stars, and coral mortality accelerated.[2] Peruvian anchovy populations have repeatedly declined after successive El Niños, each time recovering, but (because of commercial fishing pressures) each time at lower and lower levels.[3] The carcasses of fur seals littered the beaches of Point San Juan, Peru, in 1983 when a densely populated seal colony declined by 33 percent, primarily owing to pup mortality.

Climatological drift—be it on an interannual, decadal, or longer scale—necessarily exerts evolutionary pressure. When established patterns of temperature, precipitation, snow cover, or soil moisture change, the playing field upon which species compete ceases to be level. Species that are better adapted to the induced change will survive at the expense of others that are not so well suited.[4] Australia's red kangaroo retards the in utero growth of fertilized eggs during that continent's dry El Niño years, somehow knowing that better times will come later. The Galapagos penguin does not attempt to breed when sea surface temperatures rise above 24°C. Its neighbor, the waved albatross of Española Island, simply walked off the job in late 1982, abandoning eggs as El Niño conditions became more severe. Flightless cormorants chose to practice family planning when times got tough in early 1983 but were back in the saddle as soon as the water cooled and reproductive prospects improved at the end of July.[5]

"Guano" birds, such as Peruvian boobies and guanay cormorants, rely on fish that normally thrive in the cold upwelling waters off Peru. A survey in 1981–82 revealed as many as 5,750,000 birds. Only a year later, 3,700,000 adult birds were dead or had disappeared, devastated by warm El Niño waters that drove away the anchovies. By 1984–85, all surveyed nests held healthy young chicks, and a year later their numbers had climbed back to 3,160,000.[6] Adaptation in the guano birds' case is the ability to rebound after a hard year. To survive as a species, these animals have learned to cope with the environmental roller coaster that exists in the equatorial Pacific. The study of their behavior and adaptations offers a biological tracer of climates in the past.

The biological effects of climate variability can be felt around the world. El Niño conditions pushed the thermocline of the California Current down by 50 m in 1983 and raised Pacific Ocean surface temperatures by 2–4°C off the western coast of the United States.[7] Phytoplankton populations plummeted, Coho salmon remained gaunt, and sea lions starved. Increased moisture in the American Southwest during the prolonged El Niño of the early 1990s was associated with an abundance of seed grasses, an explosion of the deer mouse population, and the emergence of hantavirus as a clearly lethal pathogen.[8]

Humans have had to learn to cope. Between 1980 and 1999, the United States suffered forty-four weather-related disasters that bore a total cost of $210 billion. The Pacific Northwest drought in 1988, Hurricane Andrew in Florida in 1992, the Red River Valley floods in North Dakota in 1997: The list goes on and on.[9] In the first eleven months of 1998 alone, insurance companies worldwide documented that losses from storms, floods, droughts, and fire reached $98 billion; by Novem-

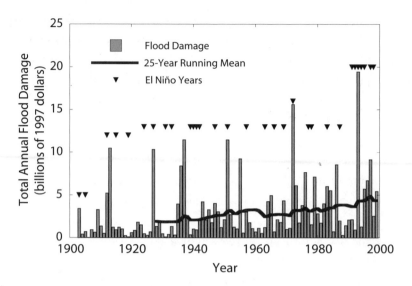

Flood damage in the United States, 1903–99 (in billions of 1997 dollars). The twenty-five-year running mean has doubled in the past seventy-five years. Spikes tend to correlate with El Niño years. (Based on Pielke and Downton 2000)

ber, 32,000 people had died in weather-related disasters.[10] In the twentieth century, yearly damages from floods rose from $43 million to $3.2 billion per year.[11]

For millennia, the Chinese have contended with annual floods on the Yangtze River. Averaging 34,000 m³/s at its mouth, the river transports 549 million tonnes of silt each year away from the Tibetan Plateau and the northern provinces of China. Over the ages, this pattern of flooding and alluviation had built up a fertile basin that now produces at least half of the country's rice. Rain fell on the basin in 1998: as much as 1,730 mm at Quinzhou in Guangxi Province during just June and July. The Yangtze rose up as the dragon it is, flushing 240 million people from their homes. Seven discrete flood crests swept down the river. One crest peaked just 20 cm below the level at which government officials had planned to intentionally flood out 330,000 more victims.[12]

Even before this flood, the Chinese had already committed themselves to a variety of flood-control projects. Preliminary work had begun on the Three Gorges Dam on the Yangtze in 1994; when finished, this $29 billion lump of concrete will create a lake 663 km long. It will have drowned 13 cities, 140 towns, and 1,352 villages and will have driven 1.2 million people from their homes. In return, the dam will generate 18,200 megawatts of electricity and presumably stop or minimize flooding on the lower Yangtze River.[13] But massive flood-control projects are not the only way to avert disasters. After the 1998 flood, the Chinese government admitted that flooding had been greatly magnified by land-management practices in the upper Yangtze basin. Logging on the Tibetan Plateau had reduced forest cover to 18 percent of its natural state. The resulting soil erosion had devastated 110,000 km^2 of Sichuan Province alone. In August 1998, the Chinese government banned logging in provinces throughout Tibet. Animals adapt to the vagaries of climate. We cope.

Watching the World Warm

With the industrial revolution in the 1800s, the human race began a high-stakes experiment: Can we alter our environment in such a way that we trigger an unstable response from the earth? For the first two hundred years, we didn't even know that we were engaged in this experiment. Only in the last few decades has the magnitude of our gamble become apparent. More was at stake than anyone had realized, and more than some are now willing to admit.

Greenhouse Gases

Charles Keeling first started to track atmospheric CO_2 in 1953 when he worked in a geochemistry lab at the California Institute of Technology. Contemplating the acidity of lakes, he speculated that a particular piece of common scientific wisdom was wrong: Carbon dioxide in the air is *not* in equilibrium with carbon dioxide in water. Propelled by the momentum of the International Geophysical Year, he began to collect atmospheric CO_2 samples from stations in Antarctica, Hawaii, and elsewhere.

Keeling already knew that CO_2 was asymmetrically partitioned between air and water: 2.6×10^{12} tonnes in the atmosphere, and 130×10^{12} tonnes in the oceans. As his early experiments got underway, he found that CO_2 concentrations varied rhythmically between winter and summer. This was no great surprise because each year plants throughout the

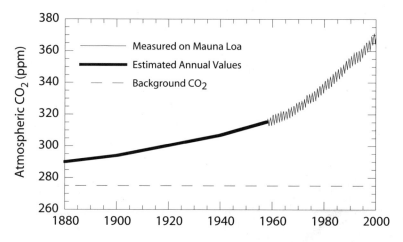

Atmospheric carbon dioxide concentrations, 1880–2000. (Based on Keeling and Whorf 1995)

world produce 200 billion tonnes of organic matter;[1] significantly more than half of this is produced on continents in the Northern Hemisphere. Because the basic building block of this productivity is atmospheric CO_2, it made sense that atmospheric concentrations would fluctuate seasonally.

What Keeling wasn't prepared for was the emerging realization that, when the annual signal was subtracted, his instruments demonstrated an inexorable upward trend in atmospheric CO_2 concentrations. He noted an interesting but as yet unexplained decadal oscillation overprinted on the data, perhaps suggesting an interaction between atmospheric CO_2 and El Niño or El Niño–like processes. The overall trend, however, was definitely upward.[2] In March 1958, he measured 314 parts per million (ppm) on Mauna Loa; thirty years later, he measured 350 ppm, same time of year, same station.[3] Current CO_2 levels in the atmosphere reflect a 30 percent rise over pre–Industrial Age values, which averaged about 280 ppm.[4]

A significant fraction of this CO_2 can be attributed to wholesale deforestation around the world, thus crippling the biological machinery necessary for the gas's extraction from the atmosphere. A larger fraction of the rise is due to widespread industrialization. In 1860 the burning of coal released 93 million tonnes of carbon into the atmosphere. In 1958 we burned gas, oil, and coal at an annual rate that released 2.3 billion tonnes. By the

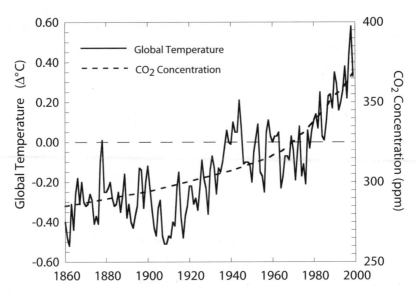

The relationship between global mean temperature and atmospheric carbon dioxide concentration, 1857–1997. (Based on Trenberth 1998)

end of the 1980s, that figure had more than doubled to 5 billion tonnes of carbon per year, and it continues to grow by an additional 4 percent each year. It turns out that Keeling was right: CO_2 does not quickly achieve an air/water equilibrium. We now know that as much as 40 percent remains in the atmosphere for centuries after it is released.[5]

So what does this portend? Let's return to our discussion of the earth's energy balance for a moment. Most solar energy arrives in wavelengths between 0.4 and 0.7 μm. About half of this is actually absorbed by the earth. This energy, in turn, is reradiated in wavelengths between 3 and 30 μm, with most in the 8–15 μm spectrum. This radiation would escape back into space, except that CO_2 is particularly adept at absorbing energy in the 13–17 μm range. As CO_2 concentrations rise, more and more of the upward energy flow is absorbed and thus retained by the climate system. The earth's lower atmosphere is undeniably warming. Scientists agree that the earth's mean temperature at the surface has risen by at least 0.5°C during the last century; many believe that the figure is now closer to 0.8°C.[6] Realistic predictions suggest that 1.0 to 3.5°C of additional warming will occur by the year 2100.

Global Warming

You may be able personally to handle a degree here, another degree there, as the earth heats up. Perhaps it means breaking out the handkerchief and mopping your brow a few more times a day than before. But can the ice sheets of Antarctica and Greenland handle it? Climate models suggest that if the low to midlatitudes heat by 2°C, the poles will simultaneously heat by 8–10°C. The ice that blankets western Antarctica, unlike the ice on the eastern portion of that continent, is grounded on rocks that lie, on average, 440 m below sea level.

Like the ice in western Antarctica, the seaward extension of the Columbia Glacier rests on the floor of Alaska's Prince William Sound. Columbia's tidewater front had been protected for centuries by the shallows adjacent to Heather Island. But a retreat that started in 1979 turned catastrophic as the glacier's terminus receded from the island's protection. Glaciologists have since concluded that Alaskan glaciers grounded in deep water are inherently unstable as they begin to retreat.[7] They have also noticed that the majority of Alaska's one hundred thousand glaciers are thinning or retreating.[8]

Are there similarities between Alaska and Antarctica? If (or is it *when?*) the 3.3 million km^3 of West Antarctic ice starts to melt, will it disappear in thousands or in hundreds of years?[9] In the unlikely event that all of Alaska's glaciers were to melt completely, sea level would rise no more than 30 cm.[10] But if West Antarctica turns out to be unstable in a recessional environment, world sea level would rise 8 m.[11] Is there a threshold atmospheric or oceanic temperature at which this potentially unstable melting would be initiated?[12] The people of Bangladesh, whose country is mostly at or near sea level, are very interested to know.

The earth has certainly seen rapid reorganizations of atmospheric patterns in the past. Evidence for these turnovers can be found within ice cores taken from Greenland and Antarctica. Some are thought to have been triggered by the shrinking of glacial ice sheets at the end of the Pleistocene. The sudden influx of freshwater would have overwhelmed the oceans' thermohaline circulation, resulting in a catastrophic collapse of this global mechanism of heat redistribution. Atmospheric effects would quickly follow.

The next century's predicted 1–3.5°C temperature escalation does not exceed warmings already well documented in the paleoclimatic record books, but never has the temperature risen so fast.[13] How do we grapple with this specter of sudden global warming, if it *is* so sudden and unusual? Should we dismiss it as a political ruse perpetrated by environmental Luddites?

Perhaps it's possible to adopt a scientist's sanguine objectivity and view anthropogenic climate change as just another interesting atmospheric behavior. This dispassion becomes more difficult, of course, when you realize that no single species has ever unleashed such abrupt change on the earth before. It is unnerving to look over your shoulder and see how many lives are at stake. Michael Dettinger, a USGS hydrologist at the Scripps Institution of Oceanography, clearly expresses the fear that global warming holds an immediate threat to our way of life when he says, "That click you just heard was the cocking of the gun that's held against our head." A high-stakes experiment, indeed.

Prediction and Modeling

The weather station at the airport in Payson, Arizona, consists of a rock suspended beneath a crude wooden tripod. Unlike other reports that have come across the wire over the years, the interpretive sign nailed to one of the tripod legs has never misled me.

If the rock is moving, it's windy.
If the rock is wet, it's raining.
If the rock is sticking straight out, you shouldn't fly.

Weather prediction is a tricky business. Those glib television forecasters in plaid pants and bow ties surely have one of the world's highest suicide rates. Their job is among the few in which one can be wrong more than half the time and not get fired. In our society, however, we still shoot the messengers. The lot of the climatologist is only slightly better.

When dire predictions of a wet winter were broadcast throughout California in advance of the 1997–98 El Niño, that state's jaundiced population laughed all the way to the beach from the warm December into January. Then it started to rain. Some $550 million in damages later, climatologist Kelly Redmond observed, "People don't believe forecasts until things have already begun to happen."

Scientists have developed a wide array of mathematical models to predict what the atmosphere will look like in a month, a year, or ten years. These models attempt to integrate vast fields of oceanographic and atmospheric data—both historical and real-time observations—and spin them into the future. The computer capacity required to run the models is astonishing. The fundamental components of the models include grids with thousands of point measurements of solar radiation, ocean currents, global wind patterns, and sea surface temperatures. All of these ingredients, of course, are already interacting with themselves. Then more volatile variables are thrown into the brew: air temperature, soil moisture, snow cover, atmospheric vapor content, and partial pressures of various atmospheric constituents. Real-time meteorologic observations are regularly integrated into the computer runs as reality checks, poured in like extra quarts of oil to lubricate the models as they whir and grind.

Global warming throws a difficult curveball to climate modelers, one with which they have no prior experience.[14] It is one thing to model the trajectory of a weather system for a few days starting from a detailed initial observation. It's quite another to predict long-term climatic changes based on changes that are based on other changes. We just don't know how all the variables will interact. For instance, increased sea-surface temperatures will increase evaporation and thus may increase cloud cover. More incoming solar energy would then be reflected back into space, but more outgoing earth radiation would be reabsorbed by clouds. Hotter or colder, wetter or drier? We just don't know.

Global circulation models are getting better at reconstructing the past.

That's a start. Eugene Rasmusson of NOAA's Climate Analysis Center, considering yet another muddled forecast back in 1986, stated, "We're still waiting for that wonderful day when a model forecasts something before it occurs."[15] Perhaps we shouldn't expect perfect predictions. Instead, it may be enough just to understand the process. "Now-casting"— that is, understanding the processes of weather as it occurs—has become a reasonable short-term cottage industry for meteorologists.

Today it is possible to monitor the eastward migration of warm water from the western Pacific Ocean and make meaningful predications about El Niño events that are likely to occur weeks to months later. ENSO is important not because it explains everything or dominates the variability in our climate. In fact, ENSO explains only about 30 percent of year-to-year climate variability in the Northern Hemisphere as a whole and locally may explain far less. ENSO is important because it is a significant component of the climate machine that we are beginning to explain, monitor, and, to some degree, predict. Integrating a new understanding of atmospheric behaviors such as the Pacific Decadal Oscillation will further refine these forecasts.

Who will benefit from these forecasts? Farmers could adjust their crops to withstand anticipated wet or dry conditions. The power industry could hedge its bets against how hot or cold an upcoming season would be. In September 1999, the Chicago Mercantile Exchange opened a market in weather futures, making it possible for investors to bet on temperatures at selected sites around the United States.[16] You can bet your bottom dollar that commodities brokers have already turned into climate forecasters as they try to squeeze just a little more juice out of their Florida orange futures. As often happens in an information-driven economy, people with access to data may fare better than others. It will be interesting to see if those who live nearest to the barrel of this climatological gun—Peruvian fishermen or Honduran farmers, for example—will have the room to move one way or the other when they see the hammer coming down.

The world is a wondrous and perplexing machine. Scientists, struggling to understand climate, are only now assembling an indispensable network of buoys, barometers, and anemometers to monitor real-time data about the oceans and atmosphere. The models will improve. But no matter how fast the computers, no matter how clever the climate models, the projections are only as good as the thinking that went into the programming in the first place. Their benefits will be realized only if we are willing to act in accordance with believable forecasts produced by the models.[17]

Making the Connection 14

$\int$cientists with NOAA document meteorologic and oceanographic parameters not just in the United States but throughout the world. The USGS measures river levels and monitors individual floods and droughts. This ongoing accumulation of data is an essential tool for meeting the day-to-day demands placed upon the nation's hydrologic resources. Knowing what we do about the coupled nature of oceans and atmosphere, we are now on the brink of doing so much more than mere monitoring. We can "see" Kelvin waves rolling toward Peru months before teleconnected El Niño events crash into North America. It is now possible — indeed, necessary — to step back and take a larger integrated view of floods and droughts. They do not happen in isolation; they occur within the context of climate.

We humans are an antlike species, working so hard with our heads tucked down, rolling stones uphill. At the beginning of the twentieth century, farmers and land sharks saw a shining potential for agriculture in southern California. They tapped the Colorado River near Yuma in an effort to turn some of its water toward irrigation projects in the Imperial Valley. This project got underway at what turned out to be one of the wettest decades in the river's recorded history. The floor of the Imperial Valley lies 200 feet below sea level. The Colorado knew a good thing when it saw one; that tap turned into torrent as the annual spring flood of 1905 breached the canal's head works. Within a year, the lower Imperial Valley, once dry, became today's Salton Sea. The Southern Pacific Railroad spent the next two years dumping trainload after trainload of rock into the breach before finally forcing the river back into its southbound channel.[1]

As a technology-based society, we are still trying to push those stones uphill. Most of our grand-scale enterprises are attempts to mitigate the impact of climatic variability. If only drought didn't threaten the Southwest, we wouldn't need a complicated system of dams and canals throughout the Colorado River drainage, and we wouldn't need to mine groundwater so feverishly from beneath Tucson and Las Vegas. If only flooding didn't threaten the Mississippi River basin, we wouldn't need 5,600 km of levees on the main stem and its tributaries.[2] We invest in these massive projects as a form of insurance, allowing us to live a little closer to the river's edge, allowing a few more people to migrate to the desert, allowing a few more homes to crowd the gulf coastline. We build these projects based on cultural memories of floods and droughts.

But how much insurance does history suggest that we buy? It is no longer sufficient to know the average flood or the average drought. In landscapes shaped by extremely wet events, the concept of the hundred-year flood loses relevance with each new ENSO event. In a world that is suddenly warming, it's difficult to know which rules still govern this game. The extremes of climate must be integrated into engineering and design thinking; otherwise the unexpected will inevitably defeat our systems of water supply, flood control, and drought mitigation. Even then, what do we have?: a technology-based insurance system that allows the world to shoehorn a few hundred million more people into harm's way.

In 1983 the U.S. Bureau of Reclamation was taken by surprise as an unexpectedly heavy spring runoff rushed down the Colorado River following the El Niño of 1982–83. In June, Lake Powell rose 15 cm a day behind Glen Canyon Dam. The bureau was unable to safely release water from the lake as fast as it was arriving. The rising water threatened to spill over the top of the 200-m-high dam. In a desperate move, plywood boards were thrown across the dam's spillways, holding back the top 2 m of a 300-km-long lake. It worked, at least that time. Now in spring, the engineers leave a little more room in the lake, just in case.

If we choose to be shortsighted, we can slap this newfound ability to decipher climate up against the wall of the future, just like those plywood boards atop Glen Canyon Dam. We can exploit these ideas to extend the effectiveness of dikes and dams right up to their breaking point, raising the ante as we go. We can adjust agricultural practices year by year to cope with drought in this time of global warming. We can employ long- and short-range forecasting to squeeze the last measure of protection out of the flood control systems upon which we have come to rely. We can continue to dump trainloads of boulders into our mistakes when the rivers run high.

Perhaps we should think twice, though, before doing so. As the world's population continues to grow, the rhythm of flood and drought is transformed from a natural climatic pattern to a series of crises that threaten more and more people. Despite all of our machinations, climate-related impacts incurred tolls in 1998 that were 50 percent higher than those of any previous year. As more humans populate the world, we find ourselves increasingly dependent on both hard and soft systems of disaster aversion. And we leave ourselves with progressively less room to maneuver within the reality of environmental variability.

The study of climate is fraught with so many statistical shadings and so many variables. It's easy to become disoriented, stalled by the uncertainties, but it would be a great loss not to move ahead with what we do know. Certainly we should continue to explore this new understanding of a globally connected climate and integrate that knowledge into the fabric of our lives. Certainly we should critically reevaluate the burning of fossil fuels, logging of rain forests, and other human activities that accelerate the processes of global warming. Perhaps one day the Peruvian fishermen in Salaverry will appreciate that their ocean's temperature is somehow connected to the floods that rush down from the Andes. Perhaps one day farmers in southeastern Alberta will be able to plant their crops with more confidence. The potential benefits of this knowledge are immense. Will we use that knowledge wisely?

Notes

Chapter 1. Entering the Atmosphere

1. Watson, L., 1984, *Heaven's Breath*: New York, William Morrow, 384 p.

Chapter 2. Cutting Edges of Climate

1. Larson, E., 1999, *Isaac's Storm*: New York, Crown Publishers, 323 p.

2. Barry, J.M., 1997, *Rising Tide*: New York, Simon and Schuster, 524 p.

3. Burroughs, W.J., 1997, *Does the Weather Really Matter?*: Cambridge, Cambridge University Press, 230 p.

4. Burroughs, W.J., 1992, *Weather Cycles: Real or Imaginary?*: Cambridge, Cambridge University Press, 201 p.

5. Kahya, E., and Dracup, J.A., 1993, U.S. streamflow patterns in relation to the El Niño/Southern Oscillation: *Water Resources Research*, v. 29, no. 8, p. 2491–2503.

6. Namias, J., 1955, Some meteorological aspects of drought, with special reference to the summers of 1942–54 over the United States: *Monthly Weather Review*, v. 83, p. 199–205.

7. Nemanishen, W., 1998, *Drought in the Palliser Triangle*: Calgary, Prairie Farm Rehabilitation Administration, 58 p.

8. Gorman, J., 1988, *A Land Reclaimed*: Hanna, Alberta, Gorman and Gorman, 183 p.

9. Jones, D.C., 1991, *Empire of Dust*: Calgary, University of Alberta Press, 316 p.

10. Lopez, B., 1978, *River Notes*: Kansas City, Andrews and McMeel, 100 p.

11. Tannehill, I.R., 1947, *Drought, Its Causes and Effects*: Newark, N.J., Princeton University Press, 264 p.

12. Smith, K., and Ward, R., 1998, *Floods*: New York, John Wiley and Sons, 382 p.

13. Baker, V.R., and Costa, J.E., 1987, Flood power, *in* Mayer, L., and Nash, D. (eds.), *Catastrophic Flooding*: Boston, Allen and Unwin, p. 1–21.

14. Sieh, K., and LeVay, S., 1998, *The Earth in Turmoil*: New York, W.H. Freeman, 324 p.

15. Smith and Ward, 1998.

16. Hirschboeck, K.K., 1987, Catastrophic flooding and atmospheric circulation anomalies, *in* Mayer, L., and Nash, D. (eds.), *Catastrophic Flooding*: Boston, Allen and Unwin, p. 23–56.

17. Rodenhuis, D.R., 1996, The weather that led to the flood, *in* Changnon, S.A (ed.), *The Great Flood of 1993*: Boulder, Colo., Westview Press, p. 29–51.

18. Koellner, W.H., 1996, The flood's hydrology, *in* Changnon, S.A (ed.), *The Great Flood of 1993*: Boulder, Colo., Westview Press, p. 68–100.

19. Changnon, S.A., 1996, Defining the flood: A chronology of key events, *in* Changnon, S.A (ed.), *The Great Flood of 1993*: Boulder, Colo., Westview Press, p. 3–28.

Chapter 3. The Weather Machine

1. Hidore, J.J., and Oliver, J.E., 1993, *Climatology, an Atmospheric Science*: New York, Macmillan, 423 p.

2. Kiehl, J.T., and Trenberth, K.E., 1997, Earth's annual global mean energy budget: *Bulletin of the American Meteorological Society*, v. 78, no. 2, p. 197–208.

3. Budyko, M., 1986, *The Evolution of the Biosphere*: Dordrecht, The Netherlands, D. Reidel, 423 p.

4. Pinet, P.R., 1998, *Invitation to Oceanography*: Sudbury, Mass., Janes and Bartlett, 508 p.

Chapter 4. The Geology of Climate

1. Tarbuck, E.J., and Lutgens, F.K., 1992, *The Earth*: New York, Macmillan, 654 p.

2. Capricorn Press Pty, 1985, *Antarctica*: Surry Hills, Mass., Reader's Digest Services, 320 p.

3. Ruddiman, W.F., and Prell, W.L., 1997, Introduction to the uplift-climate connection, *in* Ruddiman, W.F. (ed.), *Tectonic Uplift and Climate Change*: New York, Plenum Press, p. 3–15.

4. Molnar, P., England, P., and Martinod, J., 1993, Mantle dynamics, uplift of the Tibetan Plateau, and the Indian monsoon: *Review of Geophysics*, v. 31, no. 4, p. 357–396.

5. Huber, N.K., 1989, *The Geologic Story of Yosemite National Park*: U.S. Geological Survey Bulletin 1595, 64 p.

6. Press, F., and Siever, R., 1998, *Understanding Earth*: New York, W.H. Freeman, 682 p.

7. Alvarez, L.W., Alvarez, W., Asaro, F., and Michel, H.V., 1980, Extraterrestrial cause for the Cretaceous-Tertiary extinctions: *Science*, v. 208, p. 1095–1108.

8. Sadler, J.P., and Grattan, J.P., 1999, Volcanoes as agents of past environmental change: *Global and Planetary Change*, v. 21, p. 181–196.

9. Rampino, M.R., 1989, Distant effects of the Tambora eruption of April 1815, an eyewitness account: *EOS*, v. 70, no. 51, p. 1559–1560.

10. Budyko, M., 1999, Climate catastrophes: *Global and Planetary Change*, v. 20, p. 281–288.

11. Shaw, H.R., and Moore, J.G., 1988, Magmatic heat and the El Niño cycle: *EOS*, v. 69, no. 45, p. 1553, 1564–1565.

12. Handler, P., 1986, Possible association between the climatic effects of stratospheric aerosols and sea surface temperatures in the eastern tropical Pacific Ocean: *Journal of Climatology*, v. 6, p. 31–41.

13. Handler, P., and Andsager, K., 1990, Volcanic aerosols, El Niño, and the Southern Oscillation: *Journal of Climatology*, v. 10, p. 413–424.

14. Landsberg, H.E., and Albert, J.M., 1974, The summer of 1816 and volcanism: *Weatherwise*, v. 27, p. 63–66.

15. Mass, C.F., and Portman, D.A., 1989, Major volcanic eruptions and climate: *Journal of Climate*, v. 2, p. 566–593.

16. Simarski, L.T., 1992, *Volcanism and Climate Change*: American Geophysical Union Special Report, 27 p.

17. Dettinger, M.A., and Cayan, D.R., 1995, Large-scale atmospheric forcing of recent trends toward early snowmelt runoff in California: *Journal of Climate*, v. 8, p. 606–623.

18. Cayan, D.R., Redmond, K.T., and Riddle, L.G., 1999, ENSO and hydrologic extremes in the western United States: *Journal of Climate*, v. 12, no. 9, p. 2881–2893.

19. Redmond, K.T., and Koch, R.W., 1991, Surface climate and streamflow variability in the western United States and their relationship to large-scale circulation indices: *Water Resources Research*, v. 27, no. 9, p. 2382–2399.

20. Kahya and Dracup, 1993.

21. Reynolds, R., Dettinger, M.D., Cayan, D.R., Stephens, D., Highland, L., and Wilson, R., 1997, Effects of El Niño on streamflow, lake level, and landslide potential: U.S. Geological Survey (http://geochange.er.usgs.gov/sw/changes/natural/elnino/).

Chapter 5. Winds of Change

1. Karl, T.R., Knight, R.W., Easterling, D.R., and Quayle, R.G., 1995, Trends in U.S. climate during the twentieth century: *Consequences*, v. 1, no. 1, p. 3–12.

2. Emiliani, C., 1966, Isotopic paleotemperatures: *Science*, v. 154, no. 3751, p. 851–857.

3. An, Z., 2000, The history and variability of the East Asian paleomonsoon climate: *Quaternary Science Reviews*, v. 19, p. 171–187.

4. Huggett, R.J., 1991, *Climate, Earth Processes, and Earth History*: Berlin, Springer-Verlag, 281 p.

5. Newell, N.E., Newell, R.E., Hsiung, J., and Zhongxiang, W., 1989, Global marine temperature variation and the solar magnetic cycle: *Geophysical Research Letters*, v. 16, p. 311–314; Labitzke, K., and van Loon, H., 1988, Associations between the 11-year solar cycle, the QBO, and the atmosphere: *Journal of Atmospheric and Terrestrial Physics*, v. 50, p. 197–206; Vines, R.G., 1984, Rainfall patterns in the eastern United States: *Climate Change*, v. 6, p. 79–98.

6. Huggett, 1991.

7. Grove, J.M., and Switsur, R., 1994, Glacial geological evidence for the Medieval Warm Period: *Climate Change*, v. 26, no. 2/3, p. 143–170.

8. Grove, J.M., 1996, The century time-scale, *in* Driver, T.S., and Chapman, G.P. (eds.), *Time-Scales and Environmental Change*: London, Routledge Press, p. 39–87.

9. Dean, J.S., 1996, Demography, environment, and subsistence stress, *in* Tainter, J.A., and Tainter, B.B. (eds.), *Evolving Complexity and Environmental Risk in the Prehistoric Southwest*: Reading, Mass., Addison Wesley, p. 25–56.

10. Eden Foundation, 1996, Desertification, a threat to the Sahel: (http://www.eden-foundation.org/project/desertif.html).

11. Nicholson, S.E., 1995, Variability of African rainfall on interannual and decadal time scales, *in Natural Climate Variability on Decade-to-Century Time Scales*: National Research Council, p. 32–43.

12. Danish, K.W., 1995, International environmental law and the "bottom-up" approach: A review of the Desertification Convention, Indiana University School of Law (http://www.law.indiana.edu/glsj/vol3/no1/danish.html).

13. Gornitz, V., 1987, Climatic consequences of anthropogenic vegetation changes from 1880–1890, *in* Rampino, M.R., Sanders, J.E., Newman, W.S., and Konigsson, L.K. (eds.), *Climate*: New York, Van Nostrand Reinhold, 588 p.

Chapter 6. Oceans and Air

1. Philander, S.G.H., Gu, D., Halpern, D., Lambert, G., Lau, N.C., Li, T., and Pacanowski, R.C., 1996, Why the ITCZ is mostly north of the equator: *Journal of Climate*, v. 9, p. 2958–2972.

2. Macdonald, A.M., and Wunsch, C., 1996, An estimate of global ocean circulation and heat fluxes: *Nature*, v. 382, p. 436–439.

3. Stocker, T.F., 2000, Past and future reorganizations in the climate system: *Quaternary Science Reviews*, v. 19, p. 301–319.

4. Ebbesmeyer, C., 1998, Beachcombers' alert: Beachcombers' and Oceanographers' International Association (http://www.beachcombers.org).

5. Ingraham, W.J., and Miyihara, R.K., 1988, *Ocean Surface Current Simulations in the North Pacific Ocean and Bering Sea (OSCURS Numerical Model)*: NOAA Technical Memorandum, NMFS F/NWC-130, 155 p.

6. Miller, A.J., and Schneider, N., 1998, Interpreting the observed patterns of Pacific Ocean decadal variations, *in* Holloway, G., Muller, P., and Henderson, D. (eds.), *Biotic Impacts of Extratropical Climate Variability in the Pacific*: Manoa, Hawaii, University of Hawaii, Proceedings of Aha Huliko'a Hawaiian Winter Workshop, p. 19–27.

7. Latif, M., and Barnett, T.P., 1994, Causes of decadal climate variability over the North Pacific and North America: *Science*, v. 266, p. 634–637.

8. Pinet, 1998.

9. Broecker, W.S., 1997, Thermohaline circulation, the Achilles heel of our climate system: *Science*, v. 278, p. 1582–1588.

10. Broecker, W.S., 1995, Chaotic climate: *Scientific American*, v. 273, no. 11, p. 62–68.

Chapter 7. Climate in Hindsight

1. Quinn, W.H., 1992, A study of Southern Oscillation–related climatic activity for A.D. 622–1900 incorporating Nile River flood data, *in* Bradley, R.S., and Jones, P.D. (eds.), *Climate since A.D. 1500*: London, Routledge Press, p. 119–149.

2. Genesis 41:29–30.

3. Bradley, R.S., and Jones, P.D. (eds.), 1992, *Climate since A.D. 1500*: London, Routledge Press, 679 p.

4. Meko, D.M., 1992, Dendroclimatic evidence from the Great Plains of the

United States, *in* Bradley, R.S., and Jones, P.D. (eds.), *Climate since A.D. 1500*: London, Routledge Press, p. 312–330.

5. Dean, 1996.

6. Swetnam, T.W., and Betancourt, J.L., 1992, Temporal patterns of El Niño/Southern Oscillation—Wildfire teleconnections in the southwestern United States, *in* Diaz, H.F., and Markgraf, V. (eds.), *El Niño, Historical and Paleoclimatic Aspects of the Southern Oscillation*: Cambridge, Cambridge University Press, p. 259–270.

7. Fritts, H.C., and Shao, X.M., 1992, Mapping climate using tree-rings from western North America, *in* Bradley, R.S., and Jones, P.D. (eds.), *Climate since A.D. 1500*: London, Routledge Press, p. 269–295.

8. Thompson, L.G., Mosley-Thompson, E., and Thompson, P.A., 1992, Reconstructing interannual climate variability from tropical and subtropical ice-core records, *in* Diaz, H.F., and Markgraf, V. (eds.), *El Niño, Historical and Paleoclimatic Aspects of the Southern Oscillation*: Cambridge, Cambridge University Press, p. 295–322.

9. Thompson, L.G., 1992, Ice core evidence from Peru and China, *in* Bradley, R.S., and Jones, P.D. (eds.), *Climate since A.D. 1500*: London, Routledge Press, p. 517–548.

10. Bradley and Jones, 1992.

11. Cronin, T., Willard, D., Karlsen, A., Ishman, S., Verardo, S., McGeehin, J., Kerhin, R., Holmes, C., Colman, S., and Zimmerman, A., 2000, Climatic variability in the eastern United States over the past millennium from Chesapeake Bay sediments: *Geology*, v. 28, no. 1, p. 3–6.

12. Cole, J.E., Shen, G.T., Fairbanks, R.G., and Moore, M., 1992, Coral monitors of El Niño/Southern Oscillation dynamics across the equatorial Pacific, *in* Diaz, H.F., and Markgraf, V. (eds.), *El Niño, Historical and Paleoclimatic Aspects of the Southern Oscillation*: Cambridge, Cambridge University Press, p. 349–375.

13. Anderson, R.Y., Soutar, A., and Johnson, T.C., 1992, Long-term changes in El Niño/Southern Oscillation: Evidence from marine and lacustrine sediments, *in* Diaz, H.F., and Markgraf, V. (eds.), *El Niño, Historical and Paleoclimatic Aspects of the Southern Oscillation*: Cambridge, Cambridge University Press, p. 419–433.

14. Winograd, I.J., Coplen, T.B., Landwehr, J.M., Riggs, A.C., Ludwig, K.R., Szabo, B.J., Kolesar, P.T., and Revesz, K.M., 1992, Continuous 500,000-year climate record from vein calcite in Devils Hole, Nevada: *Science*, v. 258, p. 255–260.

15. O'Connor, J.E., Ely, L.L., Wohl, E.E., Stevens, L.E., Melis, T.S., Kale, S., and Baker, V.R., 1994, A 4,500-year record of large Colorado River floods in Grand Canyon, Arizona: *Journal of Geology*, v. 102, p. 1–9.

16. D'Arrigo, R.D., and Jacoby, G.C., 1992, A tree-ring reconstruction of New Mexico winter precipitation and its relation to El Niño/Southern Oscillation events, *in* Diaz, H.F., and Markgraf, V. (eds.), *El Niño, Historical and Paleoclimatic Aspects of the Southern Oscillation*: Cambridge, Cambridge University Press, p. 243–257.

17. Thompson, R.S., 1991, Pliocene environments and climates in the western United States: *Quaternary Science Reviews*, v. 10, p. 115–132.

18. Gagan, M.K., Ayliffe, L.K., Hopley, D., Cali, J.A., Mortimer, G.E., Chappell, J., McCulloch, M.T., and Head, M.J., 1998, Temperature and surface-ocean water balance of the mid-Holocene tropical western Pacific: *Science*, v. 279, p. 1014–1018.

19. Trenberth, K.E., 1997a, What is happening to El Niño?, *in Yearbook of Science and the Future: 1997 Encyclopaedia Britannica*, p. 88–99.

Chapter 8. The Christ Child

1. Enfield, D.B., 1989, El Niño, past and present: *Review of Geophysics*, v. 27, no. 1, p. 159–187.

2. Cane, M.A., 1983, Oceanographic events during El Niño: *Science*, v. 222, no. 4629, p. 1189–1194.

3. Graham, N.E., and White, W.B., 1988, The El Niño cycle: A natural oscillator of the Pacific Ocean–atmosphere system: *Science*, v. 240, p. 1293–1302.

4. Trenberth, K.E., 1991, General characteristics of El Niño–Southern Oscillation, *in* Glantz, M.H., Katz, R.W., and Nicholls, N. (eds.), *Teleconnections Linking Worldwide Climate Anomalies*: Cambridge, Cambridge University Press, p. 13–41.

5. Fagan, B., 1999, *Floods, Famines, and Emperors*: New York, Basic Books, 284 p.

6. Walker, G.T., and Bliss, E.W., 1932, World weather: *Royal Meteorological Society*, v. 4, no. 36, p. 53–84.

7. Glantz, M.H., 1996, *Currents of Change*: Cambridge, Cambridge University Press, 194 p.

8. Bjerknes, J., 1966a, Survey of El Niño 1957–58 in its relation to tropical Pacific meteorology: *Inter-American Tropical Tuna Commission Bulletin*, v. 12,

p. 1–62; Bjerknes, J., 1966b, A possible response of the atmospheric Hadley circulation to equatorial anomalies of ocean temperature: *Tellus*, v. 18, p. 820–829; Bjerknes, J., 1969, Atmospheric teleconnections from the equatorial Pacific: *Monthly Weather Review*, v. 97, p. 163–172.

9. Sandweiss, D.H., Richardson, J.B., Reitz, E.J., Rollins, H.B., and Maasch, K.A., 1996, Geoarchaeological evidence from Peru for a 5,000 years B.P. onset of El Niño: *Science*, v. 273, p. 1531–1533.

10. Clement, A.C., Seager, R., and Cane, M.A., 2000, Suppression of El Niño during the mid-Holocene by changes in the Earth's orbit: *Paleoceanography*, v. 15, p. 731–737.

11. Rasmusson, E.M., and Carpenter, T.C., 1982, Variations in tropical sea surface temperature and surface wind fields associated with Southern Oscillation/El Niño: *Monthly Weather Review*, v. 110, p. 354–384.

12. Rasmusson, E.M., 1984, El Niño: The ocean/atmosphere connection: *Oceanus*, v. 27, no. 2, p. 5–12.

13. Quinn, W.H., Neal, V.T., and Antunez de Mayolo, S., 1987, El Niño occurrences over the past four and a half centuries: *Journal of Geophysical Research*, v. 92, no. C13, p. 14,449–14,461.

14. Enfield, 1989.

15. Nials, F.L., Deeds, E.E., Moseley, M.E., Pozorski, S.G., Pozorski, T.G., and Feldman, R., 1979, El Niño: The catastrophic flooding of coastal Peru: *Bulletin of the Field Museum of Natural History*, v. 50, no. 7, p. 4–14, and v. 50, no. 8, p. 4–11.

16. Ortloff, C.R., Moseley, M.E., and Feldman, R.A., 1982, Hydraulic engineering aspects of the Chimu Chicama–Moche intervalley canal: *American Antiquity*, v. 47, no. 3, p. 572–595.

17. Wells, L.E., 1990, Holocene history of the El Niño phenomenon as recorded in flood sediments of northern coastal Peru: *Geology*, v. 18, no. 11, p. 1134–1137.

18. Enfield, D.B., 2001, Evolution and historical perspective of the 1997–1998 El Niño–Southern Oscillation event: *Bulletin of Marine Science*, v. 69, no. 1, p. 7–25.

19. Sun, D.Z., and Trenberth, K.E., 1998, Coordinated heat removal from the equatorial Pacific during the 1986–87 El Niño: *Geophysical Research Letters*, v. 25, no. 14, p. 2659–2662.

20. Trenberth, K.E., 1998a, El Niño and global warming: *Current*, v. 15, no. 2, p. 12–18.

21. Hoerling, M.P., Kumar, A., and Zhong, M., 1997, El Niño, La Niña, and the nonlinearity of their teleconnections: *Journal of Climate*, v. 10, p. 1769–1786.

22. Cayan, D.R., and Webb, R.H., 1992, El Niño/Southern Oscillation and streamflow in the western United States, *in* Diaz, H.F., and Markgraf, V. (eds.), *El Niño, Historical and Paleoclimatic Aspects of the Southern Oscillation*: Cambridge, Cambridge University Press, p. 29–68.

Chapter 9. Teleconnections

1. Kiladis, G.N., and Diaz, H.F., 1989, Global climatic anomalies associated with extremes in the Southern Oscillation: *Journal of Climate*, v. 2, no. 9, p. 1069–1090.

2. Glantz, M.H., Katz, R.W., and Nicholls, N. (eds.), 1991, *Teleconnections Linking Worldwide Climate Anomalies*: Cambridge, Cambridge University Press, 535 p.

3. Yarnal, B., and Diaz, H.F., 1986, Relationships between extremes of the Southern Oscillation and the winter climate of the Anglo-American Pacific Coast: *Journal of Climate*, v. 6, p. 197–219.

4. Schonher, T., and Nicholson, S.E., 1989, The relationship between California rainfall and ENSO events: *Journal of Climate*, v. 2, p. 1258–1269.

5. Storlazzi, C.D., and Griggs, G.B., 2000, Influence of El Niño–Southern Oscillation (ENSO) events on the evolution of central California's shoreline: *Geological Society of America Bulletin*, v. 112, no. 2, p. 236–249.

6. Cayan et al., 1999.

7. Webb, R.H., and Betancourt, J.L., 1992, *Climatic Variability and Flood Frequency of the Santa Cruz River, Pima County, Arizona*: U.S. Geological Survey Water-Supply Paper 2379, 40 p.

8. Andrade, E.R., and Sellers, W.D., 1987, El Niño and its effect on precipitation in Arizona and western New Mexico: *Journal of Climatology*, v. 8, p. 403–410.

9. Jetton, E.V., 1966, Stratospheric behavior associated with the southwestern cut-off low: *Journal of Applied Meteorology*, v. 5, p. 857–865.

10. Rasmusson, E.M., and Wallace, J.M., 1983, Meteorologic aspects of the El Niño/Southern Oscillation: *Science*, v. 222, p. 1195–1202.

11. Davies-Jones, R., 1995, Tornadoes: *Scientific American*, v. 273, no. 8, p. 48–57.

12. Shabbar, A., Bonsal, B., and Khandekar, M., 1997, Canadian precipita-

tion patterns associated with the Southern Oscillation: *Journal of Climate*, v. 10, p. 3016–3027.

13. Trenberth, K.E., and Hoar, T.H., 1996, The 1990–1995 El Niño–Southern Oscillation event: Longest on record: *Geophysical Research Letters*, v. 23, no. 1, p. 57–60.

14. Dai, A., and Trenberth, K.E., 1998, Global variations in drought and wet spells: 1900–1995: *Geophysical Research Letters*, v. 25, no. 17, p. 3367–3370.

15. Nicholls, N., 1985, Towards the prediction of major Australian droughts: *Australian Meteorologic Magazine*, v. 33, p. 161–166.

16. Allan, R.J., 1991, Australasia, *in* Glantz, M.H., Katz, R.W., and Nicholls, N. (eds.), *Teleconnections Linking Worldwide Climate Anomalies*: Cambridge, Cambridge University Press, p. 73–120.

17. Quinn, W.H., Zopf, D.O., Short, K.S., and Yang, R.K., 1978, Historical trends and statistics of the Southern Oscillation, El Niño, and Indonesian droughts: *Fishery Bulletin*, v. 76, no. 3, p. 663–678.

18. Krakauer, J., 1997, *Into Thin Air*: New York, Anchor Books, 378 p.

19. Barnett, T.P., Schlese, L.D., Roeckner, E., and Latif, M., 1991, The Asian snow-cover–monsoon–ENSO connection, *in* Glantz, M.H., Katz, R.W., and Nicholls, N. (eds.), *Teleconnections Linking Worldwide Climate Anomalies*: Cambridge, Cambridge University Press, p. 191–225.

20. Dettinger, M.D., Cayan, D.R., Diaz, H.F., and Meko, D., 1998, North-south precipitation patterns in western North America on interannual-to-interdecadal time scales: *Journal of Climate*, v. 11, p. 3095–3111.

21. Chu, P.S., 1991, Brazil's climate anomalies and ENSO, *in* Glantz, M.H., Katz, R.W., and Nicholls, N. (eds.), *Teleconnections Linking Worldwide Climate Anomalies*: Cambridge, Cambridge University Press, p. 43–71.

22. Rogers, J.C., 1988, Precipitation variability over the Caribbean and tropical Americas associated with the Southern Oscillation: *Journal of Climate*, v. 1, p. 172–182.

23. Enfield, D.B., and Mayer, D.A., 1997, Tropical Atlantic sea surface temperature variability and its relation to El Niño–Southern Oscillation: *Journal of Geophysical Research*, v. 102, no. C1, p. 929–945.

Chapter 10. Hurricanes

1. Sheridan, M.F., Bonnard, C., Carreno, R., Siebe, C., Strauch, W., Navarro, M., Calero, J., and Trujillo, N., 1999, 30 October 1998 rock fall/avalanche and

breakout flow of Casita Volcano, Nicaragua, triggered by Hurricane Mitch: *Landslide News*, no. 12, p. 2–4.

2. http://www.disastercenter.com/hurricmr.htrr

3. http://www.nhc.noaa.gov/pastdeadlya1.html

4. http://www.nhc.noaa.gov

5. Neumann, C.J., Jarvinen, B.R., Pike, A.C., and Elms, J.D., 1990, *Tropical Cyclones of the North Atlantic Ocean, 1871–1986 (with Storm Track Maps Updated through 1989)*: Asheville, N.C., National Climatic Data Center, Historical Climatology Series 6-2, p. 23–24.

6. Gray, W., 2000, Colorado state hurricane forecast: (http://tropical.atmos.colostate.edu/).

Chapter 11. The Slow Wheels of Change

1. Mantua, N.J., Hare, S.R., Zhang, Y., Wallace, J.M., and Francis, R.C., 1997, A Pacific interdecadal climate oscillation with impacts on salmon production: *Bulletin of the American Meteorological Society*, v. 78, no. 6, p. 1069–1079.

2. Biondi, F., Gershunov, A., and Cayan, D.R., 2001, North Pacific decadal climate variability since 1661: *Journal of Climate*, v. 14, no. 1, p. 5–10.

3. Minobe, S., 1997, A 50–70-year climatic oscillation over the North Pacific and North America: *Geophysical Research Letters*, v. 24, p. 683–686.

4. Dettinger, M.D., Cayan, D.R., McCabe, G.J., and Marengo, J.A., 2000, Multiscale streamflow variability associated with El Niño/Southern Oscillation, *in* Diaz, H.F., and Markgraf, V. (eds.), *El Niño and the Southern Oscillation: Multiscale Variability, Global and Regional Impacts*: Cambridge, Cambridge University Press, p. 113–147.

5. Rasmusson, E.M., 1991, Observational aspects of ENSO cycle teleconnections, *in* Glantz, M.H., Katz, R.W., and Nicholls, N. (eds.), *Teleconnections Linking Worldwide Climate Anomalies*: Cambridge, Cambridge University Press, p. 309–343.

6. Delworth, T.L., Manabe, S., and Stouffer, R.J., 1995, North Atlantic interdecadal variability in a coupled model, *in Natural Climate Variability on Decade-to-Century Time Scales*: Washington, D.C., National Academy Press, Climate Research Committee, p. 432–439.

7. Zhang, Y., Wallace, J.M., and Battisti, D.S., 1997, ENSO-like interdecadal variability, 1900–93: *Journal of Climate*, v. 10, p. 1004–1020.

8. Gershunov, A., and Barnett, T.P., 1998, Interdecadal modulation of ENSO

teconnections: *Bulletin of the American Meteorological Society*, v. 79, no. 12, p. 2715–2725.

9. Miller and Schneider, 1998.

10. McCabe, G.J., and Dettinger, M.D., 1999, Decadal variations in the strength of ENSO teleconnections with precipitation in the western United States: *International Journal of Climatology*, v. 19, p. 1399–1410.

11. Gershunov, A., Barnett, T.P., and Cayan, D.R., 1999, North Pacific inter-decadal oscillation seen as a factor in ENSO-related North American climate anomalies: *EOS*, v. 80, no. 3, p. 29–31.

12. Sheridan, T.E., 1995, *Arizona: A History*: Tucson, University of Arizona Press, 434 p.

13. Dettinger, M.D., and Schaefer, D.H., 1995, Decade-scale hydroclimatic forcing of ground-water levels in the central Great Basin, eastern Nevada, *in* Herrmann, R., Sidle, R.C., Back, W., and Johnson, A.I. (eds.), *Water Resources and Environmental Hazards: Emphasis on Hydrologic and Cultural Insight in the Pacific Rim*: Herndon, Va., American Water Resources Association, Annual Summer Symposium Proceedings, TPS-95-2, p. 195–204.

Chapter 12. Living in a Web of Climate

1. Cole et al., 1992.

2. Glynn, P.W., 1988, El Niño–Southern Oscillation 1982–1983: Nearshore population, community, and ecosystem responses: *Annual Review of Ecology and Systematics*, v. 19, p. 309–345.

3. Sharp, G.D., 1992, Fishery catch records, El Niño/Southern Oscillation, and longer-term climate change as inferred from fish remains in marine sediments, *in* Diaz, H.F., and Markgraf, V. (eds.), *El Niño, Historical and Paleoclimatic Aspects of the Southern Oscillation*: Cambridge, Cambridge University Press, p. 379–417.

4. Ford, M.J., 1982, *The Changing Climate: Responses of the Natural Flora and Fauna*: London, Allen and Unwin, 190 p.; McGowan, J.A., Cayan, D.R., and Dorman, L.M., 1998, Climate-ocean variability and ecosystem response in the northeast Pacific: *Science*, v. 281, p. 210–217.

5. Valle, C.A., Cruz, F., Cruz, J.B., Merlen, G., and Coulter, M.C., 1987, The impact of the 1982–1983 El Niño–Southern Oscillation on seabirds in the Galapagos Islands, Ecuador: *Journal of Geophysical Research*, v. 92, no. C13, p. 14437–14444.

6. Tovar, H., Guillen, V., and Cabrera, D., 1987, Reproduction and popula-

tion levels of Peruvian guano birds, 1980 to 1986: *Journal of Geophysical Research*, v. 92, no. C13, p. 14445–14448.

7. McGowan et al., 1998.

8. Duchin, J.S., and the Hantavirus Study Group, 1994, Hantavirus pulmonary syndrome: *New England Journal of Medicine*, v. 330, p. 949–955.

9. National Climatic Data Center, 2000, Billion dollar U.S. weather disasters, 1980–1999: http://www.ncdc.noaa.gov/ol/reports/billionz.html.

10. Trenberth, K.E., 1999, The extreme weather events of 1997 and 1998: *Consequences*, v. 5, no. 1, p. 2–15.

11. Pielke, R.A., Jr., and Downton, M.W., 2000, Precipitation and damaging floods: Trends in the United States, 1932–97: *Journal of Climate*, v. 13, p. 3625–3637; data source: http://www.nws.noaa.gov/oh/hic/flood_stats/flood_loss_time_series.html.

12. Rekenthaler, D., 1998, China floods exacerbated by man's impact on land, climate: Disaster Relief (http://www.disasterrelief.org/disasters).

13. Mufson, S., 1999, The Yangtze Dam: Feat or folly: *Washington Post*, November 9, p. A1.

Chapter 13. Watching the World Warm

1. Budyko, 1986.

2. Keeling, C.D., and Whorf, T.P., 1995, Decadal oscillations in global temperature and atmospheric carbon dioxide, *in Natural Climate Variability on Decade-to-Century Time Scales*: Washington, D.C., National Academy Press, Climate Research Committee, p. 97–110. See also Whorf, T.P., and Keeling, C.D., 1998, Rising carbon: *New Scientist*, v. 157, p. 54.

3. Weiner, J., 1990, *The Next One Hundred Years*: New York, Bantam Books, 312 p.

4. Trenberth, K.E., 1998b, The different flavors of La Niña: Review of the causes and consequences of cold events: Boulder, Colo., La Niña Summit, July 15–17, 1998 (http://www.esig.ucar.edu/lanina/report/trenberth.html).

5. Karl, T.R., Nicholls, N., and Gregory, J., 1997, The coming climate: *Scientific American*, v. 279, p. 78–83.

6. Karl, T.R., and Trenberth, K.E., 1999, The human impact on climate: *Scientific American*, v. 281, no. 6, p. 100–105.

7. Van Der Veen, C.J., 1996, Tidewater calving: *Journal of Glaciology*, v. 42, no. 141, p. 373–385.

8. Sapiano, J.J., Harrison, W.D., and Echelmeyer, K.A., 1998, Elevation, vol-

ume, and terminus changes of nine glaciers in North America: *Journal of Glaciology*, v. 44, no. 146, p. 119–135.

9. Budd, W.F., McInnes, B.J., Jenssen, D., and Smith, I.N., 1987, Modeling the response of the West Antarctic ice sheet to a climatic warming, *in* Van Der Veen, C.J., and Oerlemans, J. (eds.), *Dynamics of the West Antarctic Ice Sheet*: Boston, D. Reidel, p. 321–358.

10. Molnia, B., 2001, Glaciers of Alaska: *Alaska Geographic*, v. 28, no. 2, 112 p.

11. Schneider, D., 1997, The rising seas: *Scientific American*, v. 279, no. 3, p. 112–117.

12. Parker, D.E., Folland, C.K., and Jackson, M., 1995, Marine surface temperature: Observed variations and data requirements: *Climatic Change*, v. 31, p. 559–600; Jones, P.D., New, M., Parker, D.E., Martin, S., and Rigor, I.G., 1999, Surface air temperature and its changes over the past 150 years: *Geophysical Reviews*, v. 37, p. 173–199.

13. Crowley, T.J., 2000, Causes of climate change over the past 1,000 years: *Science*, v. 289, p. 270–277.

14. Levi, B.G., 1990, Climate modelers struggle to understand global warming: *Physics Today*, February, p. 17–19.

15. Gannon, R., 1986, Solving the puzzle of El Niño: Is this the key to long-range weather forecasting?: *Popular Science*, v. 229, no. 9, p. 82–86, 118.

16. Seabrook, J., 2000, Selling the weather: *The New Yorker*, April 3, p. 44–53.

17. Trenberth, K.E., 1997b, The use and abuse of climate models: *Nature*, v. 386, p. 131–133.

Chapter 14. Making the Connection

1. deBuys, W., and Myers, J., 1999, *Salt Dreams*: Albuquerque, University of New Mexico Press, 307 p.

2. Weiner, N., 1993, Designing the Upper Mississippi: Background briefing: http://www.backgroundbriefing.com/missipp.html.

Glossary

albedo. Amount of surface reflectivity of incoming radiation.

Aleutian Low. Persistent region of low atmospheric pressure over the Gulf of Alaska southeast of the Aleutian Islands.

cirrus. High, white, feathery cloud made of ice crystals.

climate. Long-term pattern of precipitation and temperature at a given location. *See also* **weather.**

conduction. Transfer of energy from one molecule directly to another molecule.

convection. Mass movement within a liquid or gas that redistributes heat.

Coriolis effect. Apparent curvature of motion of an object traveling in a straight line across the surface of a rotating sphere. Because of the earth's direction of rotation, objects appear to be deflected in a clockwise direction in the Northern Hemisphere and counterclockwise in the Southern Hemisphere.

cumulonimbus. Massive cloud with a dark base and significant vertical development.

current. Movement within a fluid body; named in large bodies of water according to the direction toward which it flows. A *westerly current* flows toward the west; an *easterly current* flows toward the east. *See also* **wind.**

debris flow. Downslope motion of a viscous slurry of water and poorly sorted sediments.

dendrochronology. Study of annual growth rings of trees, used for dating and paleoclimatic analysis.

desertification. Environmental degradation leading to biologic deterioration of a landscape.

drought. Persistent deficit of atmospheric moisture below long-term average conditions. *Agricultural drought* refers to drought conditions that are sufficiently persistent and intense to negatively impact cultivated crops. *Ecologic drought* refers to drought conditions that are detrimental to native plants that have previously adapted to a location's climatic conditions. *Hydrologic drought* refers to drought conditions that are sufficient to diminish surface-water supplies.

Ekman transport. Deflection of water flow to the right of the wind's direction in the Northern Hemisphere (and to the left of the wind in the Southern Hemisphere) due to the Coriolis effect.

El Niño. (1) Episodic current of warm water along the western coast of South America near the equator, which occurs every three to seven years in December. (2) Episodic warming of surface waters of the equatorial central and eastern Pacific Ocean, which is associated with loss or reversal of the Pacific trade winds and worldwide climatic anomalies. *See also* **El Niño–Southern Oscillation, La Niña,** and **Southern Oscillation.**

El Niño–Southern Oscillation (ENSO). Coupled oceanographic and atmospheric processes of El Niño that span the equatorial Pacific Ocean.

equatorial wave guide. Tendency of waves to curve toward the equator due to the Coriolis effect.

evaporation. Transformation of a liquid into a gas due to a change toward higher heat or lower pressure.

flood. Natural or artificial rise in flow volume above a channel's normal bank level. *See also* **hundred-year flood.**

flood basalt. Extensive flow of magma from multiple fissures over a large region.

geomorphology. Study of the earth's surface that relates geologic materials and structures to landforms.

global warming. Warming of the earth's surface and atmosphere in response to a natural or anthropogenic increase in effective solar radiation, decreased albedo, or increased atmospheric insulation owing to increases in trace gases such as carbon dioxide.

greenhouse effect. Tendency of the atmosphere to retain more heat radiated by the earth as energy-absorbing gaseous constituents (such as carbon dioxide and water vapor) increase.

gyre. Rotating body of water that spans an entire ocean basin on one side or the other of the equator.

Hadley cell. Vertically oriented cell of atmospheric circulation located between the equator and latitudes either 23° north or south of the equator, with winds near the surface blowing toward the equator and winds at higher altitudes blowing away from the equator.

harmattan. Dusty, dry wind blowing from the Sahara Desert.

heat capacity. A material property measured by the quantity of heat needed to raise the temperature of one gram of a substance by 1 °C.

hundred-year flood (HYF). Statistically derived magnitude of a flood that has a 1 percent chance of occurring during any given year, based on long-term averages of known flooding in a given channel.

hurricane. Large rotating storm that begins within the tropical Atlantic Ocean with winds at least 120 km/hr. Magnitudes increase from level 1 through 5 as wind velocity increases. *See also* **tropical cyclone.**

insolation. Solar radiation that reaches the earth's surface.

interference. Interaction of forces occurring at different rates. *Constructive interference* reinforces the dynamic strength of two processes; *destructive interference* diminishes the dynamic strength of two processes.

Intertropical Convergence Zone (ITCZ). Zone of low pressure oriented parallel to the equator that seasonally migrates a few degrees north or south and is remarkable for its high moisture content and unstable air.

isotherm. Contour line of equal temperature.

Kelvin wave. Packet of warm water traveling east along the equator from the western Pacific toward South America.

La Niña. Atmospheric and oceanographic condition marked by unusually warm water in the western Pacific Ocean, strong trade winds, and strong upwelling of cold water along the western coast of South America. *See also* **El Niño, El Niño–Southern Oscillation,** and **Southern Oscillation.**

latent heat. Potential energy stored within a substance.

meridional flow. North-south flow of air or water parallel to lines of longitude. *See also* **zonal flow.**

Milankovitch cycle. Cyclic changes in the distance between the earth's surface and the sun, based on the variation in the earth's elliptical orbit around the sun and in the earth's rotation about its axis.

nilometer. One of several crude devices used to measure the flooding stages of the Nile River.

North Atlantic Oscillation. Atmospheric pressure gradient measured between Iceland and the Azores.

orographic effect. Tendency of mountain ranges to deflect moving air upward, causing it to cool and release precipitation.

Pacific Decadal Oscillation (PDO). Rhythmic warming or chilling of the central northern Pacific Ocean.

plate tectonics. Theory that the earth's crust is divided into discrete plates that can move relative to one another.

Quasi-Biennial Oscillation (QBO). High-altitude equatorial wind that blows from east or west, reversing its direction every 26–30 months.

radiation. Transfer of energy from one body to another via electromagnetic waves.

Rossby wave. Reflection of a Kelvin wave as it strikes the western coast of South America.

Southern Oscillation. Atmospheric pressure gradient between Darwin, Australia, and Tahiti. The Southern Oscillation Index (SOI) is the absolute difference between these two locations; a negative value indicates El Niño conditions, and a positive value indicates La Niña conditions.

stratosphere. Relatively stable, dry portion of the atmosphere typically extending 11–50 km above the earth's surface.

teleconnection. Coupling of individual atmospheric disturbances across great distances.

thermocline. Sharp vertical temperature gradient between water masses with markedly different temperatures.

thermodynamics. Study of the relationship between heat and mechanical work.

thermohaline circulation. Oceanic currents driven by water-density gradients that are primarily based on temperature and salinity differences.

tornado. High-velocity atmospheric vortex with very low pressure at its center.

trade wind. Two belts of wind, both blowing from the east, that converge slightly toward the equator.

tropical cyclone. Large, rotating, low-pressure system with high winds that occurs within tropical latitudes. From least to most severe, tropical cyclones are called *tropical depressions, tropical storms*, and *hurricanes* over the Atlantic Ocean or *typhoons* over the Pacific Ocean.

troposphere. Portion of the atmosphere where most weather disturbances occur. It extends from the earth's surface to about 11 km above it and is marked by temperatures, pressures, and moisture levels that decrease with altitude.

tsunami. Long-period ocean wave caused by a geologic disturbance, such as a volcanic eruption or seafloor earthquakes.

varve. Sedimentary layer laid down reliably owing to annual events.

Walker Circulation. Vertically oriented, east-west flow of air over the equatorial Pacific Ocean, with easterly winds near the surface and westerly winds at higher altitudes.

weather. State of the atmosphere at any given point in time and space. *See also* **climate.**

wind. Movement of air; named according to the direction from which it blows. A *westerly wind* blows out of the west; an *easterly wind* blows out of the east. *See also* **current.**

zonal flow. East-west flow of air or water parallel to lines of latitude. *See also* **meridional flow.**

Suggested Readings

Chapter 1. Entering the Atmosphere

Watson, L., 1984, *Heaven's Breath*: New York, William Morrow, 384 p.

Chapter 2. Cutting Edges of Climate

Burroughs, W.J., 1997, *Does the Weather Really Matter?*: Cambridge, Cambridge University Press, 230 p.

Namias, J., 1955, Some meteorological aspects of drought, with special reference to the summers of 1942–54 over the United States: *Monthly Weather Review*, v. 83, p. 199–205.

Smith, K., and Ward, R., 1998, *Floods*: New York, John Wiley and Sons, 382 p.

Chapter 3. The Weather Machine

Kiehl, J.T., and Trenberth, K.E., 1997, Earth's annual global mean energy budget: *Bulletin of the American Meteorological Society*, v. 78, no. 2, p. 197–208.

Pinet, P.R., 1998, *Invitation to Oceanography*: Sudbury, Mass., Janes and Bartlett, 508 p.

Chapter 4. The Geology of Climate

Budyko, M., 1999, Climate catastrophes: *Global and Planetary Change*, v. 20, p. 281–288.

Mass, C.F., and Portman, D.A., 1989, Major volcanic eruptions and climate: *Journal of Climate*, v. 2, p. 566–593.

Pinet, P.R., 1998, *Invitation to Oceanography*: Sudbury, Mass., Janes and Bartlett, 508 p.

Redmond, K.T., and Koch, R.W., 1991, Surface climate and streamflow variability in the western United States and their relationship to large-scale circulation indices: *Water Resources Research*, v. 27, no. 9, p. 2382–2399.

Ruddiman, W.F., and Prell, W.L., 1997, Introduction to the uplift-climate connection, *in* Ruddiman, W.F. (ed.), *Tectonic Uplift and Climate Change*: New York, Plenum Press, p. 3–15.

Sadler, J.P., and Grattan, J.P., 1999, Volcanoes as agents of past environmental change: *Global and Planetary Change*, v. 21, p. 181–196.

Chapter 5. Winds of Change

Grove, J.M., 1996, The century time-scale, *in* Driver, T.S., and Chapman, G.P. (eds.), *Time-Scales and Environmental Change*: London, Routledge Press, p. 39–87.

Huggett, R.J., 1991, *Climate, Earth Processes, and Earth History*: Berlin, Springer-Verlag, 281 p.

Karl, T.R., Knight, R.W., Easterling, D.R., and Quayle, R.G., 1995, Trends in U.S. climate during the twentieth century: *Consequences*, v. 1, no. 1, p. 3–12.

Chapter 6. Oceans and Air

Broecker, W.S., 1995, Chaotic climate: *Scientific American*, v. 273, no. 11, p. 62–68.

Broecker, W.S., 1997, Thermohaline circulation, the Achilles heel of our climate system: *Science*, v. 278, p. 1582–1588.

Latif, M., and Barnett, T.P., 1994, Causes of decadal climate variability over the North Pacific and North America: *Science*, v. 266, p. 634–637.

Macdonald, A.M., and Wunsch, C., 1996, An estimate of global ocean circulation and heat fluxes: *Nature*, v. 382, p. 436–439.

Stocker, T.F., 2000, Past and future reorganizations in the climate system: *Quaternary Science Reviews*, v. 19, p. 301–319.

Chapter 7. Climate in Hindsight

Alley, R., 2000, *The Two-Mile Time Machine*: Princeton, N.J., Princeton University Press, 229 p.

Bradley, R.S., and Jones, P.D. (eds.), 1992, *Climate since A.D. 1500*: London, Routledge Press, 679 p.

Diaz, H.F., and Markgraf, V. (eds.), 1992, *El Niño, Historical and Paleoclimatic Aspects of the Southern Oscillation*: Cambridge, Cambridge University Press, 476 p.

Quinn, W.H., 1992, A study of Southern Oscillation–related climatic activity for A.D. 622–1900 incorporating Nile River flood data, *in* Bradley, R.S., and Jones, P.D. (eds.), *Climate since A.D. 1500*: London, Routledge Press, p. 119–149.

Trenberth, K.E., 1997, What is happening to El Niño?, *in Yearbook of Science and the Future: 1997 Encyclopaedia Britannica*, p. 88–99.

Chapter 8. The Christ Child

Cane, M.A., 1983, Oceanographic events during El Niño: *Science*, v. 222, no. 4629, p. 1189–1194.

Enfield, D.B., 1989, El Niño, past and present: *Review of Geophysics*, v. 27, no. 1, p. 159–187.

Fagan, B., 1999, *Floods, Famines, and Emperors*: New York, Basic Books, 284 p.

Glantz, M.H., 1996, *Currents of Change*: Cambridge, Cambridge University Press, 194 p.

Pinet, P.R., 1998, *Invitation to Oceanography*: Sudbury, Mass., Janes and Bartlett, 508 p.

Rasmusson, E.M., 1984, El Niño: The ocean/atmosphere connection: *Oceanus*, v. 27, no. 2, p. 5–12.

Trenberth, K.E., 1991, General characteristics of El Niño–Southern Oscillation, *in* Glantz, M.H., Katz, R.W., and Nicholls, N. (eds.), *Teleconnections Linking Worldwide Climate Anomalies*: Cambridge, Cambridge University Press, p. 13–41.

Trenberth, K.E., 1998, El Niño and global warming: *Current*, v. 15, no. 2, p. 12–18.

See also http://www.pmel.noaa.gov/tao/elnino/nino-home.html.

Chapter 9. Teleconnections

Cayan, D.R., and Webb, R.H., 1992, El Niño/Southern Oscillation and stream-flow in the western United States, *in* Diaz, H.F., and Markgraf, V. (eds.), *El Niño, Historical and Paleoclimatic Aspects of the Southern Oscillation*: Cambridge, Cambridge University Press, p. 29–68.

Glantz, M.H., Katz, R.W., and Nicholls, N. (eds.), 1991, *Teleconnections Linking Worldwide Climate Anomalies*: Cambridge, Cambridge University Press, 535 p.

Ropelewski, C.F., and Halpert, M.S., 1987, Global and regional scale precipitation patterns associated with the El Niño/Southern Oscillation: *Monthly Weather Review*, v. 115, p. 1606–1626.

Ropelewski, C.F., and Halpert, M.S., 1989, Precipitation patterns associated with the high index phase of the Southern Oscillation: *Journal of Climate*, v. 2, p. 268–284.

Chapter 10. Hurricanes

Dunn, G.E., and Miller, B.I., 1964, *Atlantic Hurricanes*: Baton Rouge, Louisiana State University Press, 326 p.

Sheets, B., and Williams, J., 2001, *Hurricane Watch, Forecasting the Deadliest Storms on Earth*: New York, Vintage Books, 331 p.

Chapter 11. The Slow Wheels of Change

Biondi, F., Gershunov, A., and Cayan, D.R., 2001, North Pacific decadal climate variability since 1661: *Journal of Climate*, v. 14, no. 1, p. 5–10.

Gershunov, A., Barnett, T.P., and Cayan, D.R., 1999, North Pacific interdecadal oscillation seen as a factor in ENSO-related North American climate anomalies: *EOS*, v. 80, no. 3, p. 29–31.

Mantua, N.J., Hare, S.R., Zhang, Y., Wallace, J.M., and Francis, R.C., 1997, A Pacific interdecadal climate oscillation with impacts on salmon production: *Bulletin of the American Meteorological Society*, v. 78, no. 6, p. 1069–1079.

McCabe, G.J., and Dettinger, M.D., 1999, Decadal variations in the strength of ENSO telecommunications with precipitation in the western United States: *International Journal of Climatology*, v. 19, p. 1399–1410.

Minobe, S., 1997, A 50–70-year climatic oscillation over the North Pacific and North America: *Geophysical Research Letters*, v. 24, p. 683–686.

Chapter 12. Living in a Web of Climate

Pielke, R.A., Jr., and Downton, M.W., 2000, Precipitation and damaging floods: Trends in the United States, 1932–97: *Journal of Climate*, v. 13, p. 3625–3637; data source: http://www.nws.noaa.gov/oh/hic/flood_stats/flood_loss_time_series.html.

El Niño and Biology: http://www.coaps.fsu.edu/lib/elninobib/fisheries/

El Niño and Marine Life: http://www.csa.com/hottopics/elnino/absfish.html

Chapter 13. Watching the World Warm

Crowley, T.J., 2000, Causes of climate change over the past 1,000 years: *Science*, v. 289, p. 270–277.

Karl, T.R., Nicholls, N., and Gregory, J., 1997, The coming climate: *Scientific American*, v. 279, p. 78–83.

Karl, T.R., and Trenberth, K.E., 1999, The human impact on climate: *Scientific American*, v. 281, no. 6, p. 100–105.

Keeling, C.D., and Whorf, T.P., 1995, Decadal oscillations in global temperature and atmospheric carbon dioxide, *in Natural Climate Variability on Decade-to-Century Time Scales*: Washington, D.C., National Academy Press, Climate Research Committee, p. 97–110.

Schneider, D., 1997, The rising seas: *Scientific American*, v. 279, no. 3, p. 112–117.

Trenberth, K.E., 1998, El Niño and global warming: *Current*, v. 15, no. 2, p. 12–18.

Weiner, J., 1990, *The Next One Hundred Years*: New York, Bantam Books, 312 p.

Chapter 14. Making the Connection

Floods and Droughts: http://walrus.wr.usgs.gov/elnino/

Climate Prediction: http://www.cpc.ncep.noaa.gov/

Pacific Decadal Oscillation: http://tao.atmos.washington.edu/pdo/

Index

Note: Page numbers in italics indicate illustrations.

About the Authors

Michael Collier is a science writer, photographer, and physician from Flagstaff, Arizona.

Robert H. Webb is a research hydrologist in Tucson, Arizona, who studies climate change in the southwestern United States. Both Collier and Webb are with the U.S. Geological Survey.